Factories & Office Buildings

LINKS

Factories & Ofiice Buildings
Edition 2008

Author: Carles Broto
Publisher: Carles Broto
Graphic designer & production: Raquel Castro
Text: Contributed by the architects, edited and translated by Wiliiam George and Marta Rojals.

Jonqueres, 10, 1-5
08003 Barcelona, Spain
Tel.: +34 93 301 21 99
Fax: +34 93 301 00 21
E-mail: info@linksbooks.net
www.linksbooks.net

ISBN: 978-84-96424-74-6

Factories & Office Buildings

LINKS

Index

Introduction

For many years, industrial architecture has been relegated to the outskirts of cities. Factories have been conceived as just that, with no thought given to aesthetics, making them mere containers for industrial activity. It was the interior which mattered, and the validity of the architecture lay only in whether or not cost-effectiveness was improved.

This book documents the changes going on in the field of industrial architecture. Due, in part, to the large-scale proliferation of industrial complexes, which seem to arise in our midst without care or planning, entrepreneurs, architects and public institutions are showing an increased interest in new ways of conceiving the office buildings and industrial structures of the future. Aware of the imprint made by these buildings on the landscape and of an increasing demand to create sustainable environments, architects are creating spaces which are not only worker-friendly, but which have also been conceived with respect toward their surroundings.

This volume is a collection of 24 works done by such internationally renowned architects as Krueck & Sexton, Henn Architekten, Jaques Ferrier and NL Architects, among others.

The factories, offices, warehouses, and waste treatment plants in this book are accompanied by floor plans and construction details which provide a better understanding of their composition and the work done by their creators. These are projects which show the endless possibilities offered by a reinterpretation of 21st century industrial architecture – a wide range of typologies into which each architectural studio has put its own personal touch.

Álvaro Siza

Vitra Production Building

Weil am Rhein, Germany

Photographs: Richard Bryant

In the design of Vitra´ s new industrial building for the production of prototypes, factors taken into consideration included not only functional and production needs but also the integration of the building into its environment, an industrial estate where the varied forms, scales and functions of the different premises reach an optimal combination.
While its sobriety of forma is in keeping with the deconstructivist fire station by Zaha Hadid, the great metal arm of the canopy reaches out towards a large preexisting building, clad with enamelled iron panels, by Nicholas Grimshaw.
The roof of the building, in steel with linear roof-lights, is supported by reinforced concrete columns distributed in the same way as in the other factories in the complex. All the technical installations (gas, electricity, telephone, lighting and heating) are concealed within the metal structure of the roof, thus guaranteeing maximum flexibility. On the interior, just two volumes punctuate the diaphanous spaces of this 9.000 sqm industrial unit: the control booth, and a concrete service block housing the toilets and the production manager´ s office. The circular control booth is duction on pillars and built in glass and steel to ensure maximum visibility throughout the building; from here the functioning of the factory´ s machinery is controlled by computer.
As the building stands on a plot originally used for parking, Siza also included a car ark on the western edge of the complex. A prefabricated concrete grid alternatives with asphalt and lets the grass grow, while strategically placed trees provide shade.

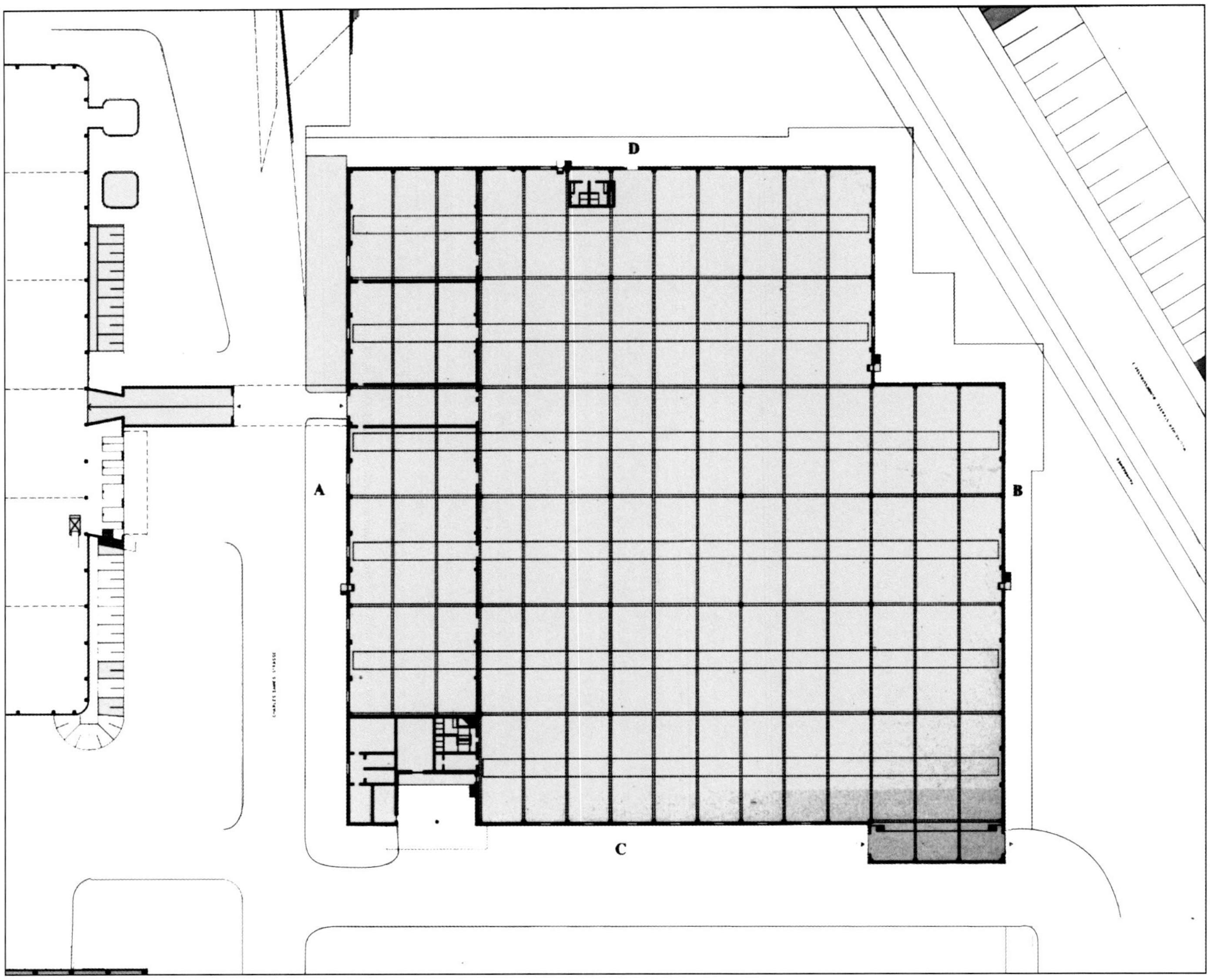
D
A
B
C

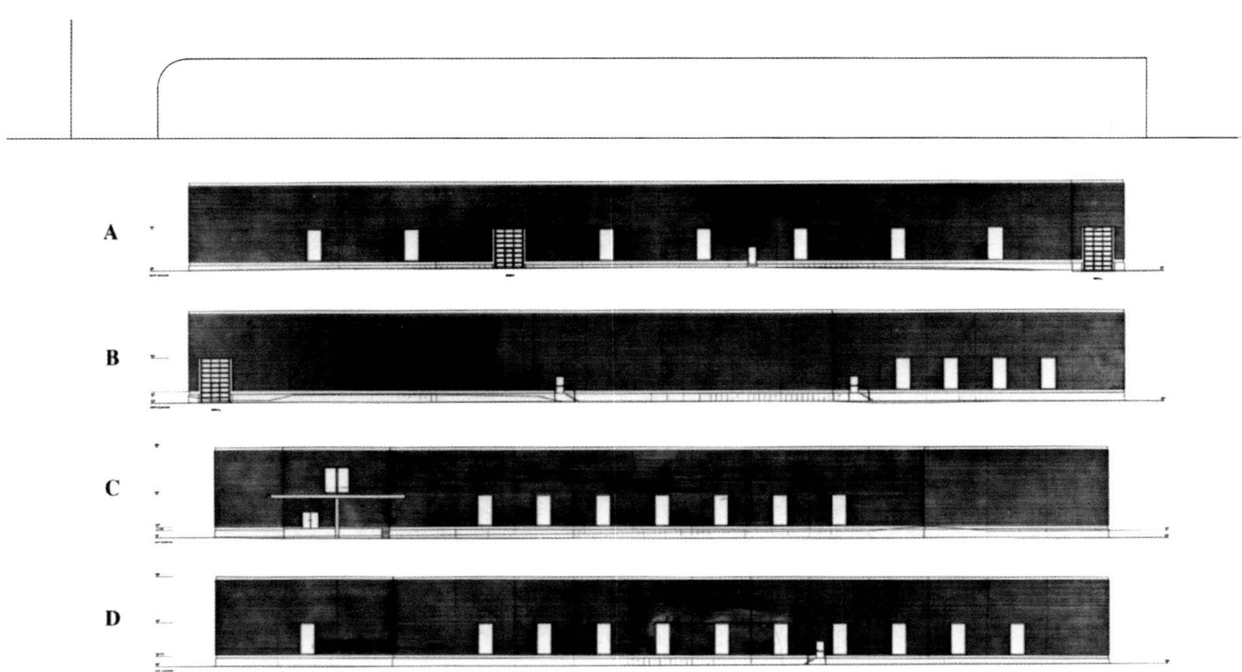
A
B
C
D

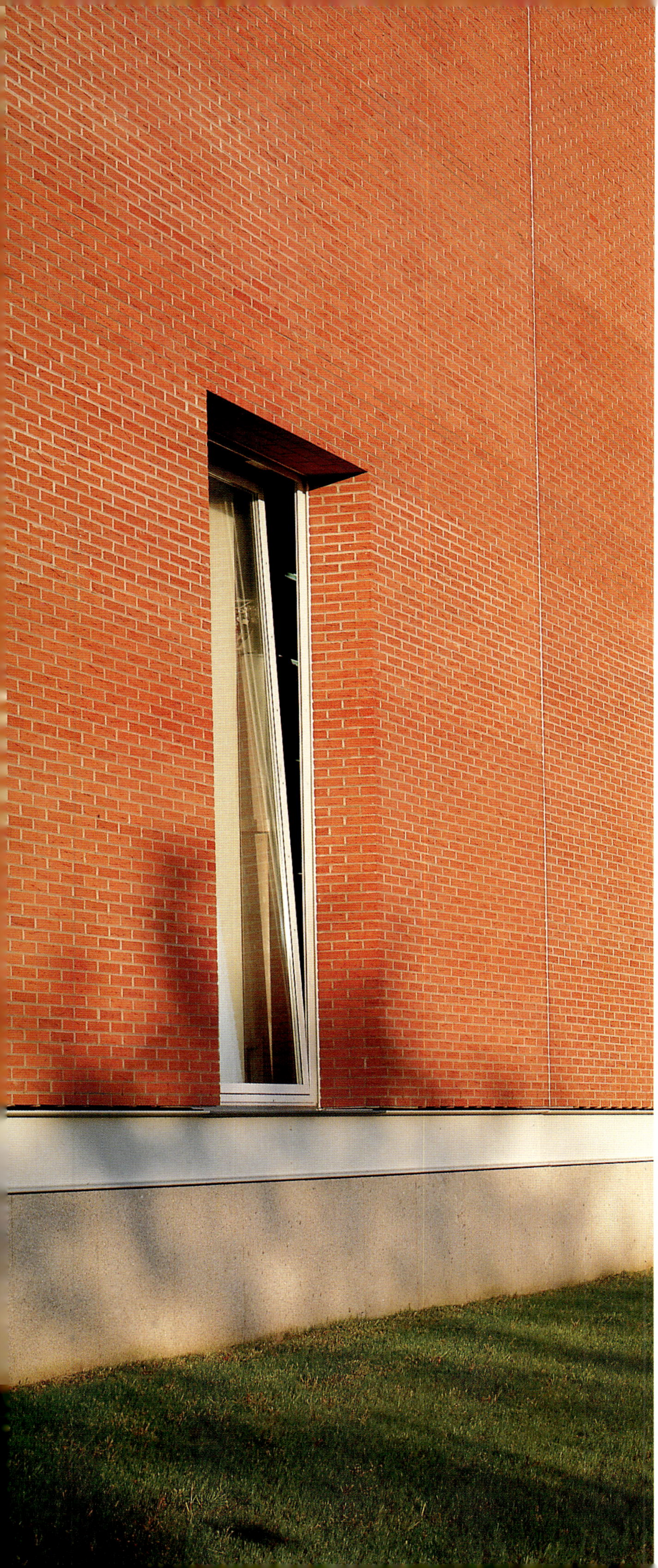

The uniformity of the extensive textured red brickwork facade is interrupted only by the tall steel-profiled Windows. White U-shaped steel profiles are also used to articulate the points of contact with the base in Portuguese granite and the cornice of the building.

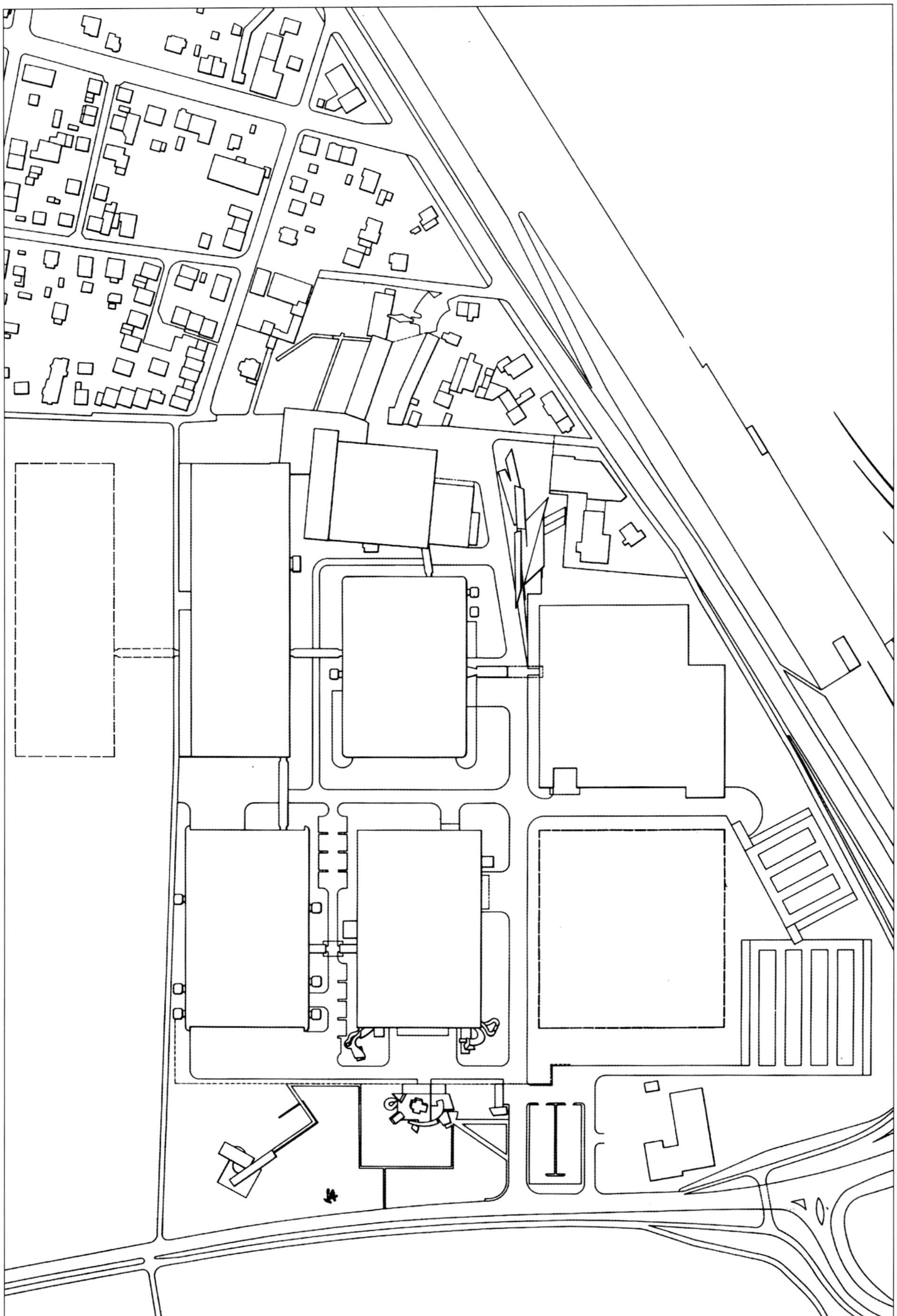

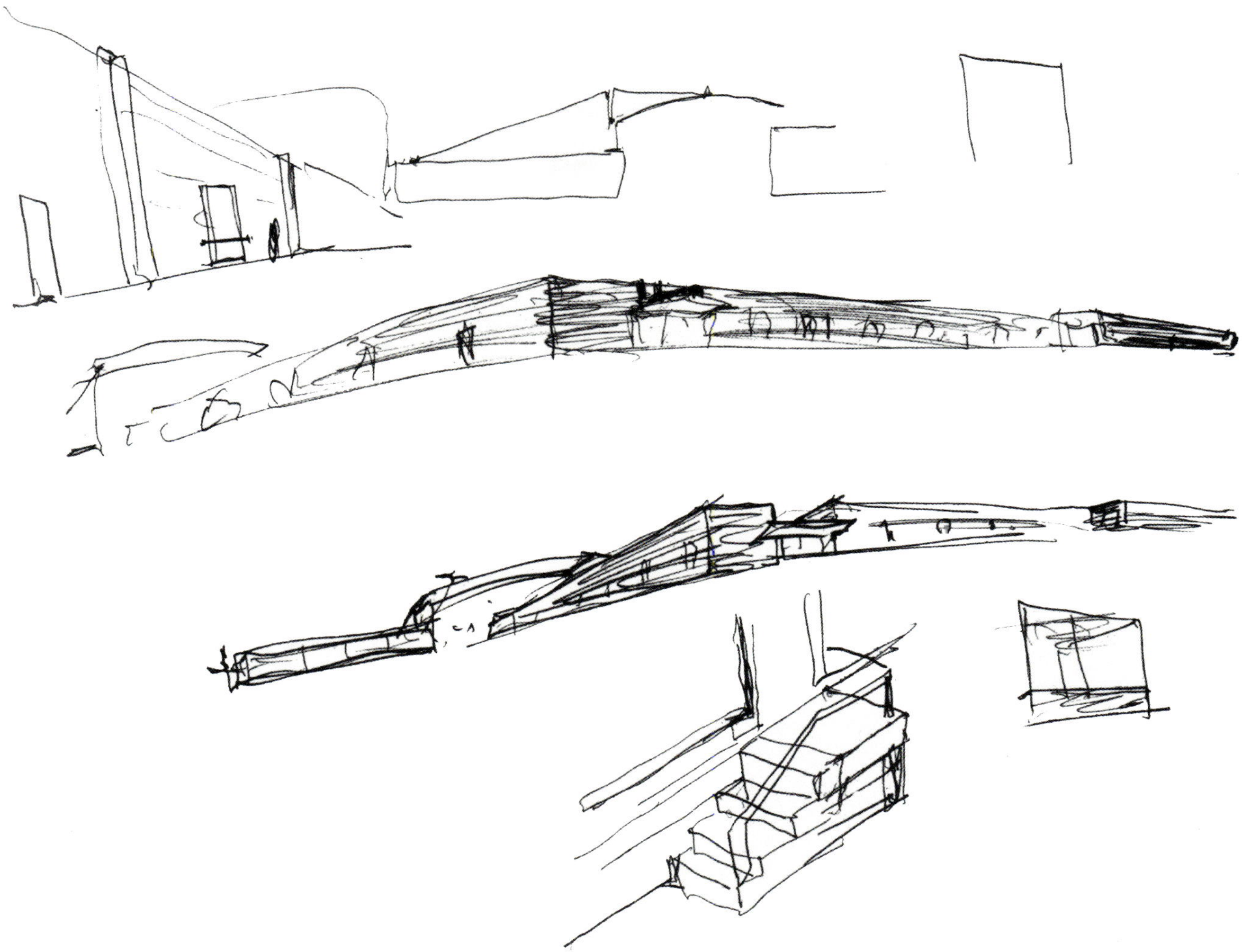

Sketches of the production hall by A.Siza and aerial photograph of the Vitra complex.

The steel canopy, suspended by tour steel cables, can be raised into an unobtrusive, slimline position or lowered to Project users and merchandise from the elements.

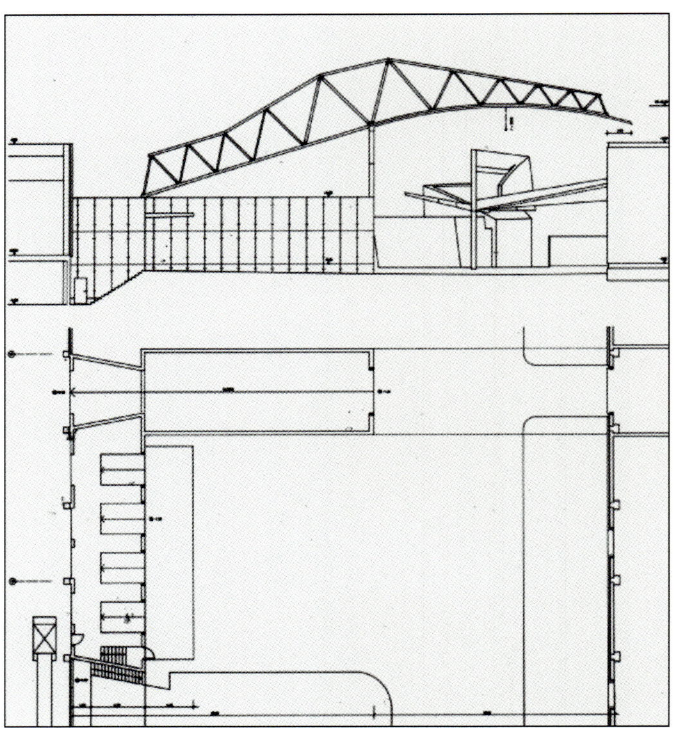

STOP
2
Vitrashop
Wareneingang 1

The control booth contains the computers that control the machinery. Its raised position and glass structure assure maximum visibility from this room.

The design of the service block, with its structural segmented roof, is base don that of the traditional houses in the region. It houses the toilets and the production manager ´s office.

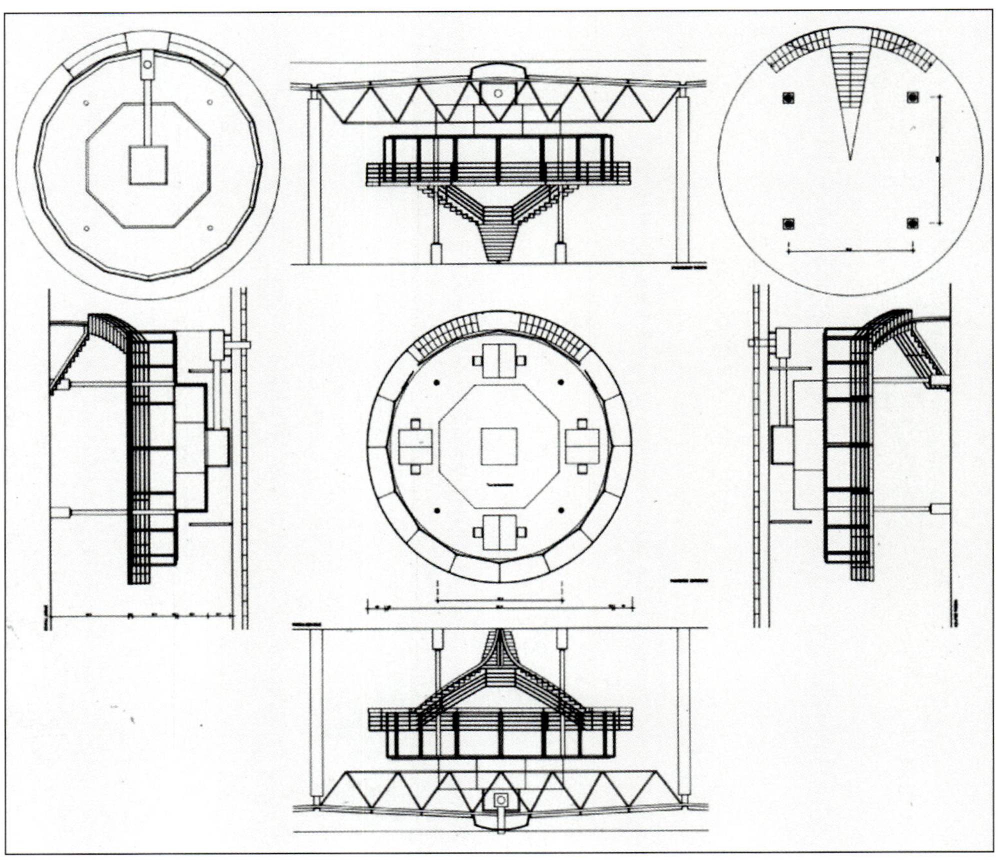

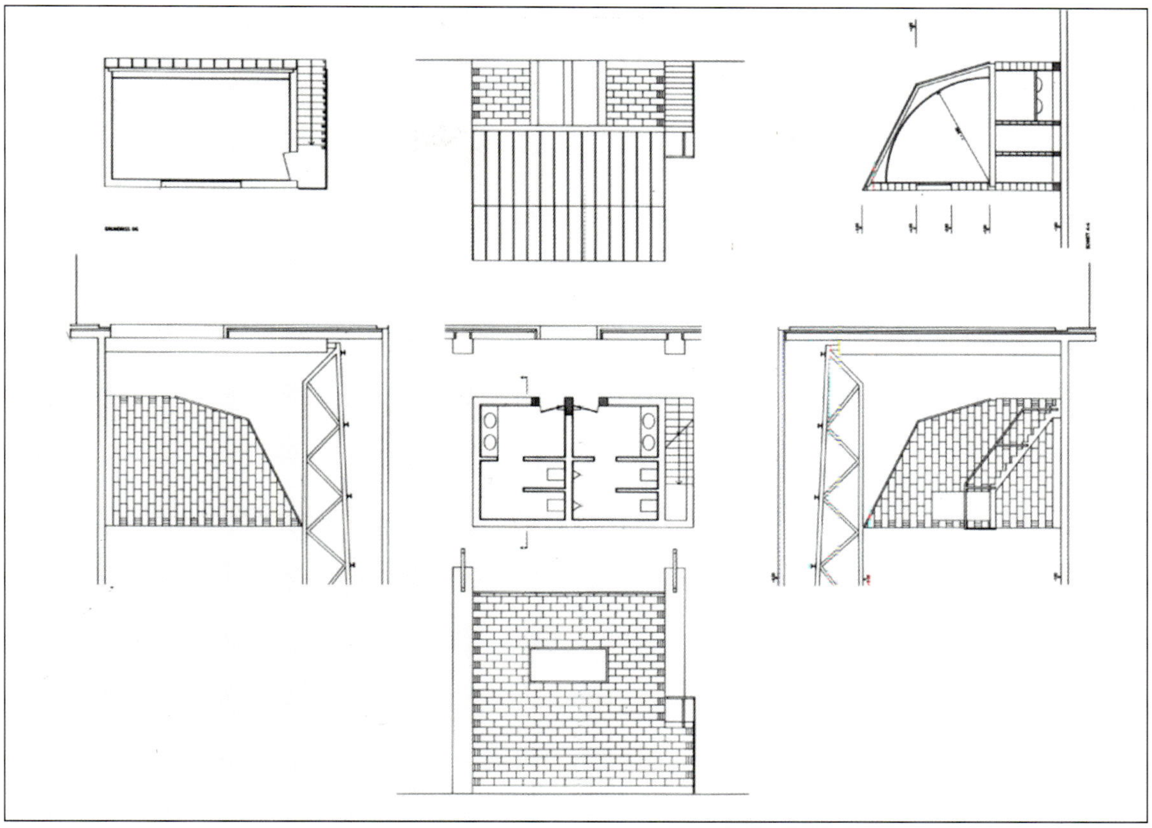

Ian Ritchie Architects

Stockley Park Building B8

London, U.K.

Photographs: Iaj Lawson

B8 is an elegant building designed with a hierarchy of scale such that each component relates proportionally to the others. The result is one of harmony, in which the eye is not disturbed yet the building is visually alive. The architects sought simplicity combined with good environmental performance and economy, and Ian Ritchie describes the building as "a pinstripe businessman who is catching a plane".

Practical completion of both shell and core of the three-storey, 7600 sqm building was required within 52 weeks of the architects´ appointment, and the cladding had to come within a budget of £300 per sqm. The solution was to apply a refinement of the Pilkington double-glazed planar system to panels approximately 3 m wide by 1.5 m, held by four pins to mullions.

A high environmental performance is achieved by the yse of ceramic fritting, low-emissivity coatings and external sunscreens for the full-height glazing. The design of the horizontal fritting allows unobstructed vision at seated and standing eye levels and is tuned to avoid interference patterns. It also serves to mask visual clutter in the floor zone,

Lightwells and roof-lights flood the internal areas with zenithal light; this and the lateral light from the main facades provide high levels of glare-free natural daylight throughout the floor spaces.

External noise had to be controlled on the roof as well as the facades due to a nearby helicopter depot. The mass of the shell was increased accordingly, and the glass walls consist of 12 mm outer and 6 mm inner panes with a 16 mm air gap.

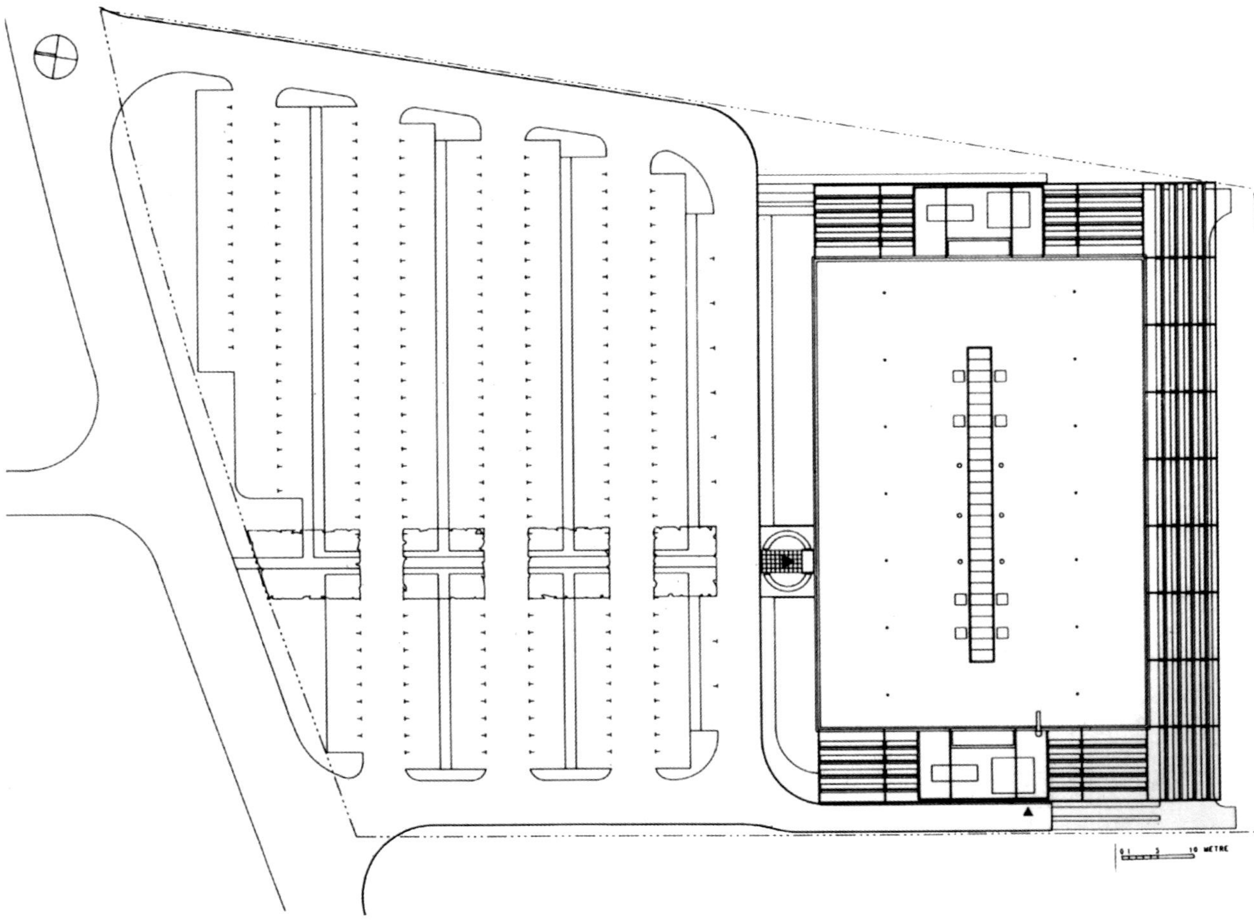

The B8 Building at Stockley Park comprises 7.600 sqm of net office accommodation arranged three floors to a 9 by 9 m construction grid. Two 18 by 63 m clear office floor plates flank a central service core and lightwell providing for single or multi-tenancy occupation.

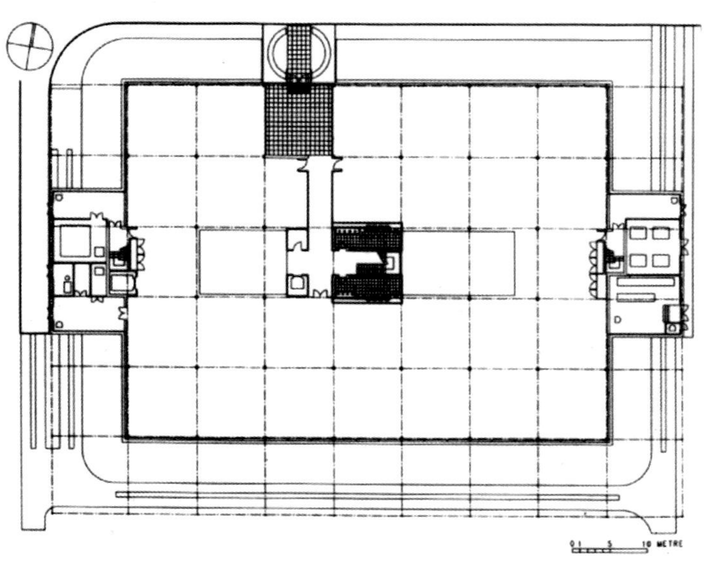

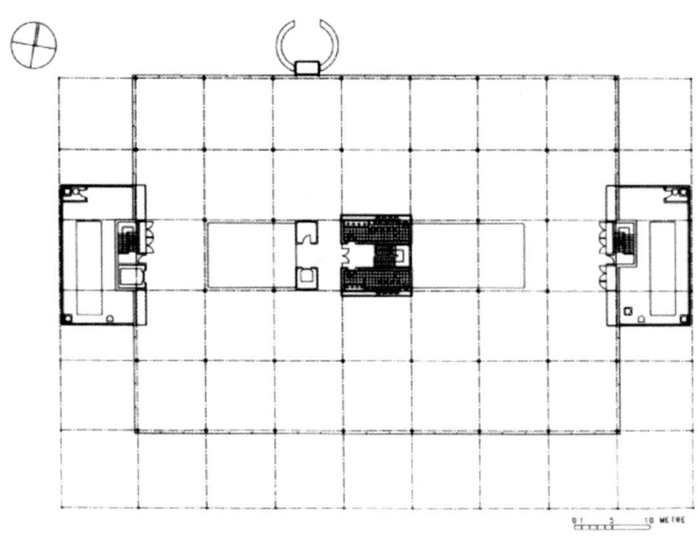

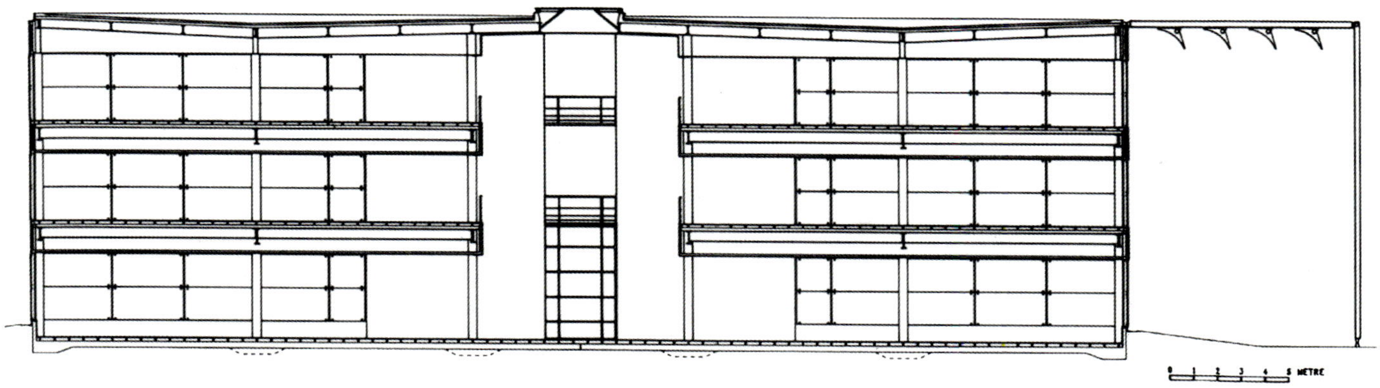

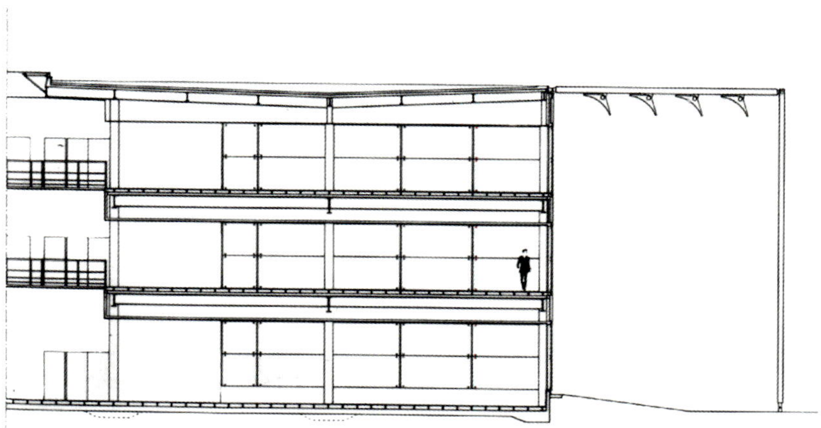

Above: sections. Below: detail of the sun screens, present on all the facades except the north.

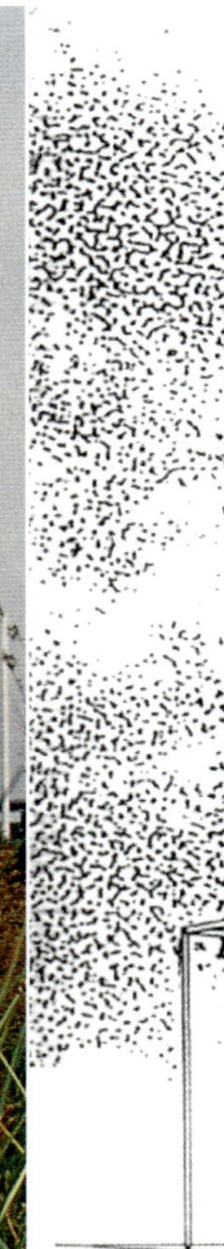

The finished building provides a discreet, efficiently organized, high-technology facility. Below: north and south facades.

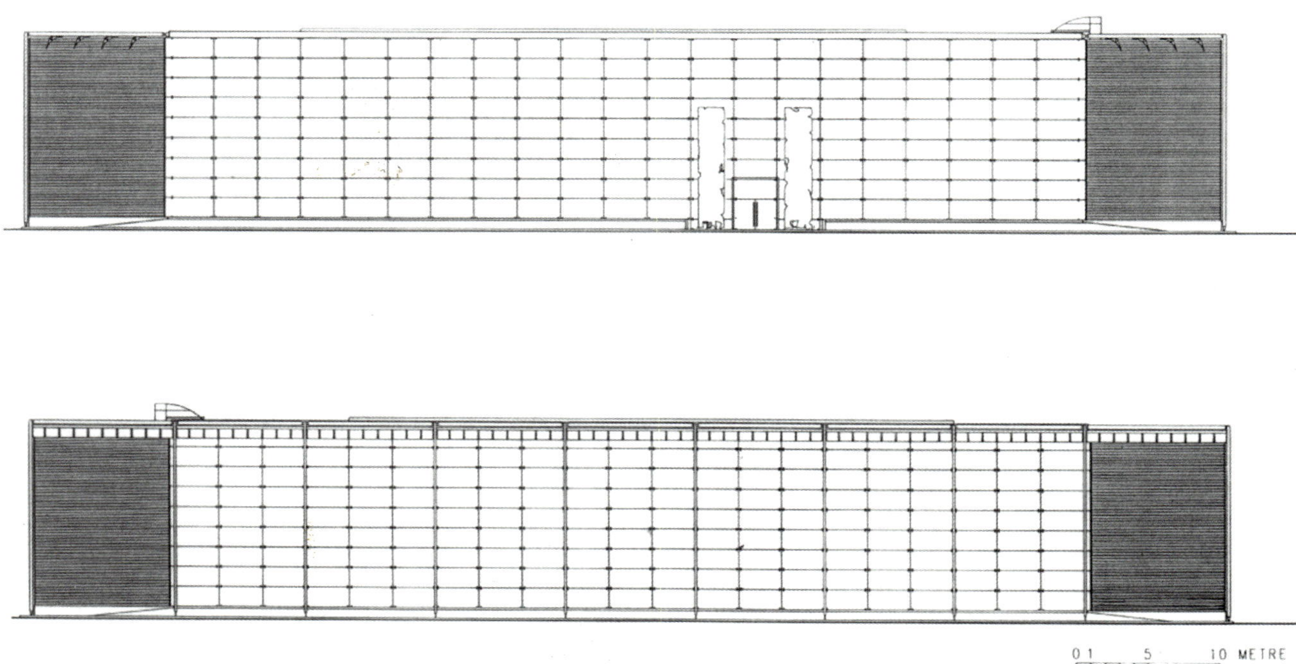

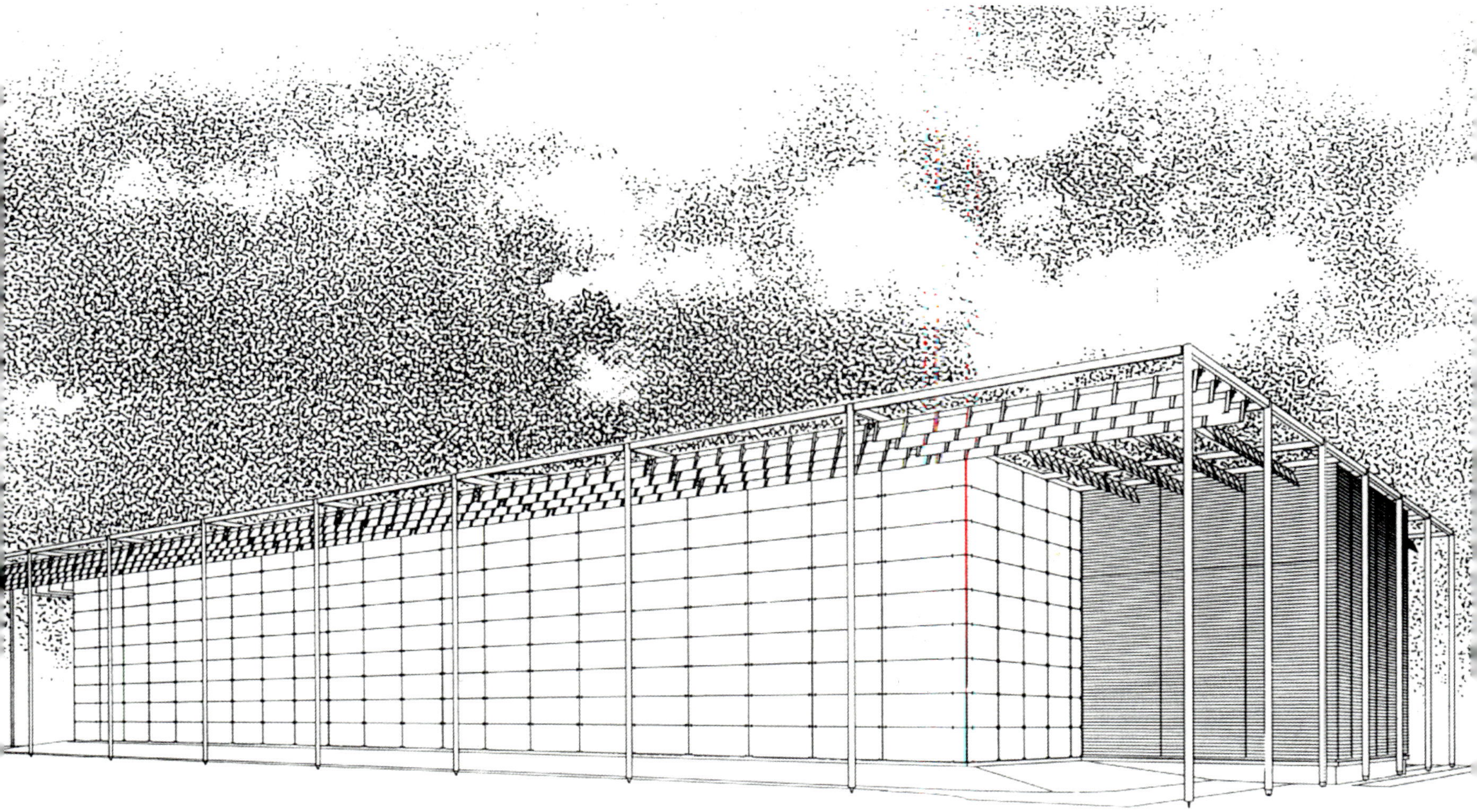

Above: perspective. Below: east and best facades.

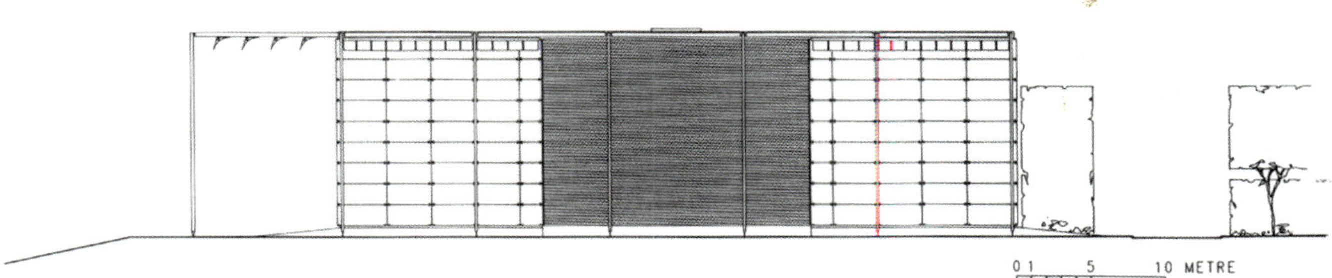

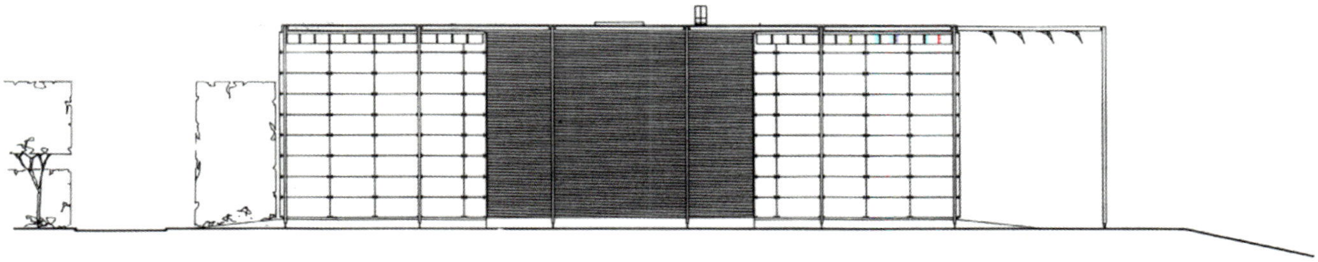

The building grid is extended by a full bay to support large curved perforated stainless steel sun screens on the south, east and best facades.

The pattern of Light and shade on the glass during the course of the day results in a continually changing modulations of the flan grid of the building face.

Typical bay on the east, best and south elevations. The design of the fritting has carefully avoided the "spots-before-the-eyes" syndrome and stereoscopic interference.

Below: the fritted areas, living 70% opacity, are surprisingly transparent from inside. Clear visual areas the normal range of eye heights allow the occuoant to take the panoramic view.

The characteristic greenish colour of the panel is produced by the thickness of the glass. Black silicone was for the panel junctions and the double-glazed units to ensure compatibility.

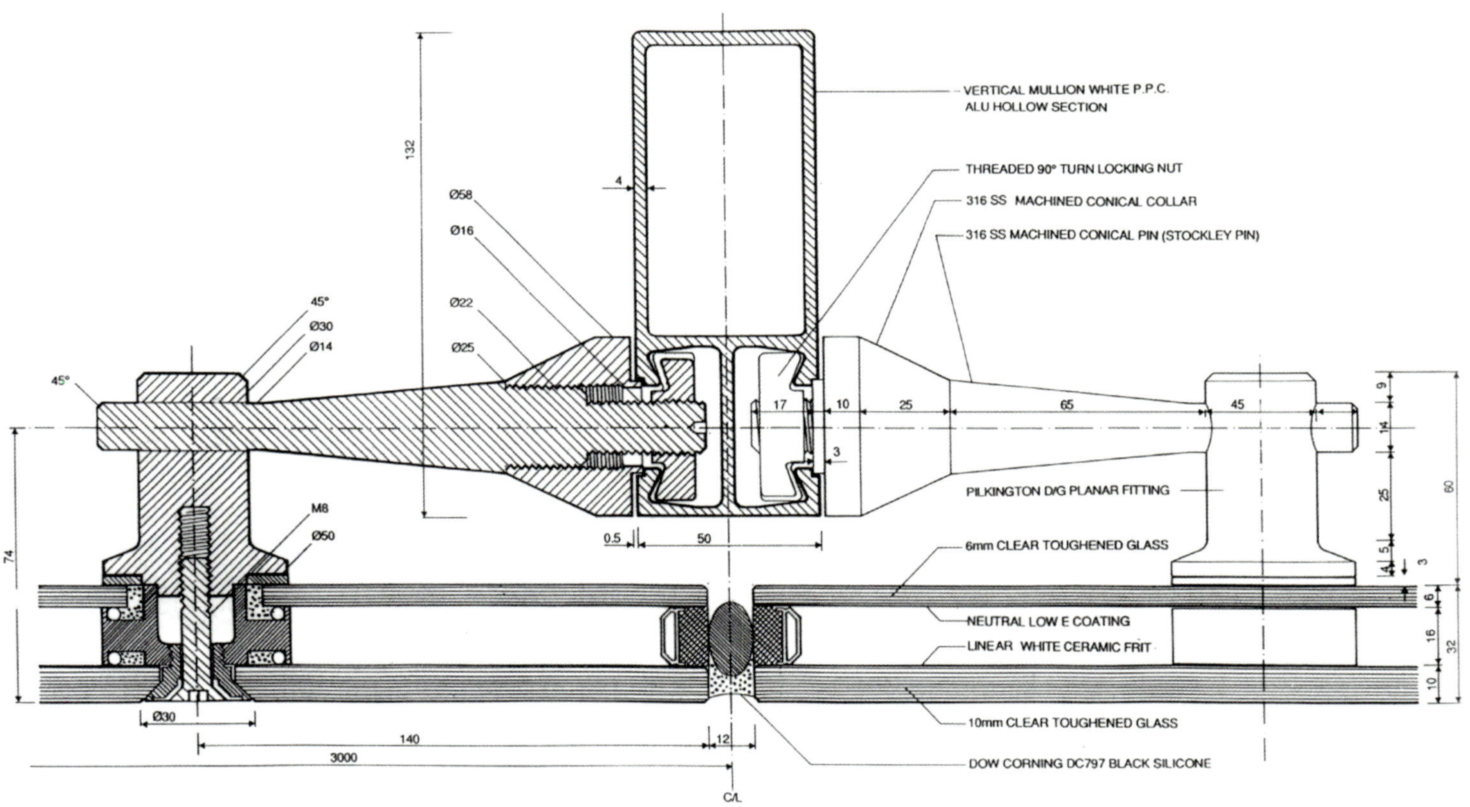

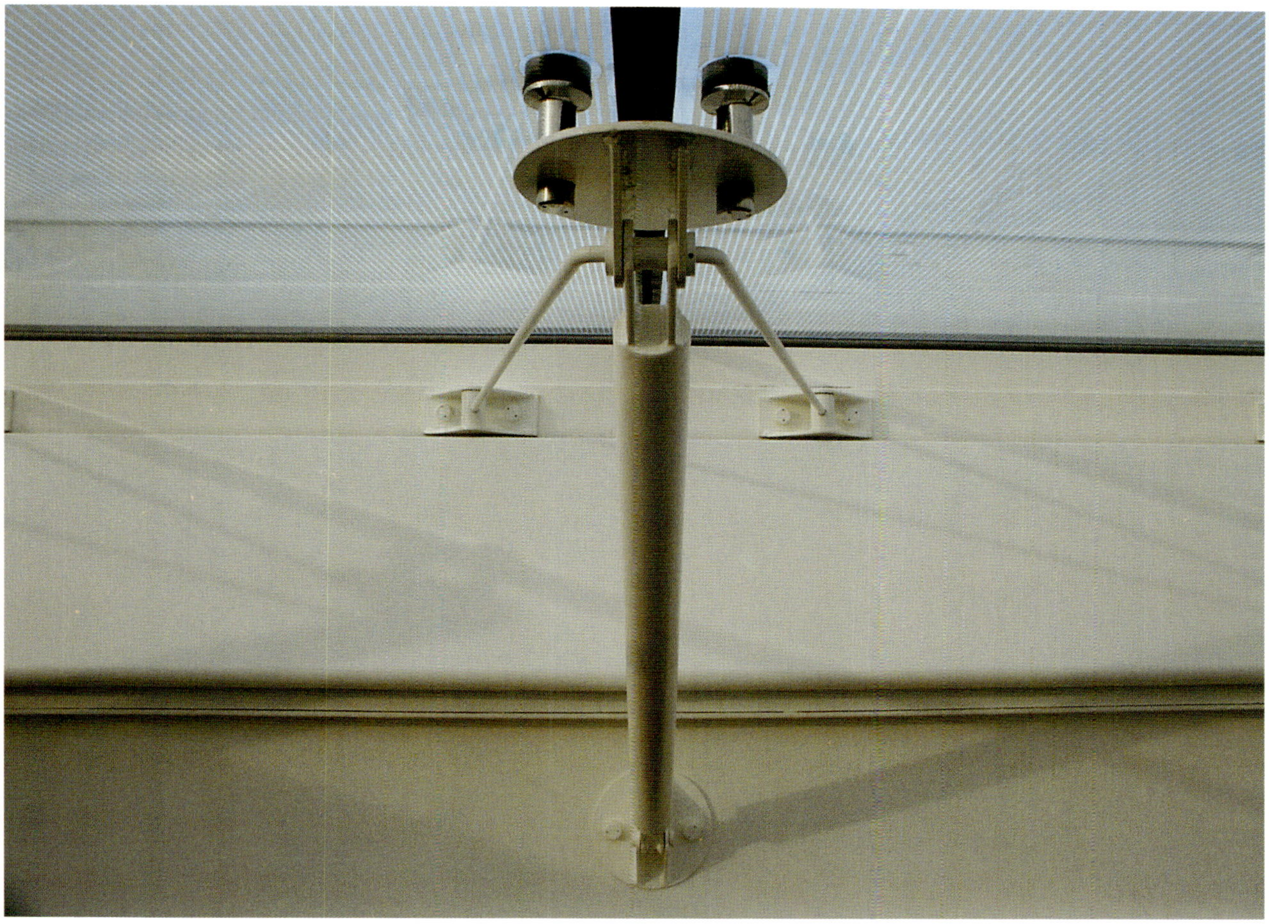

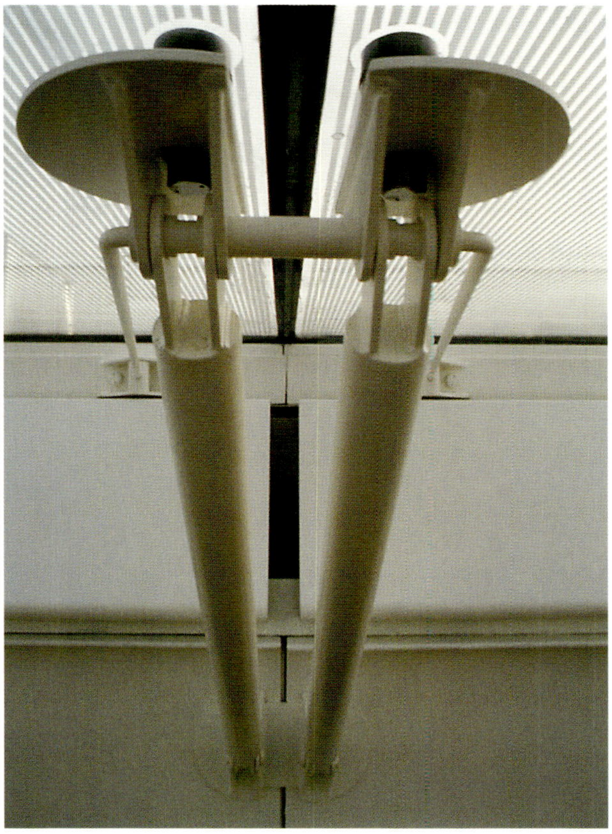

The aluminium mullion extrusions are specially designed to receive stainless steel support pins in a continuous groove which provide vertical fixing tolerances, while the pin design accommodates horizontal tolerances and movement. The full-height mullions are supported at ground level and restrained at each floor for wind loads. The famous Ove Arup team realised the structural design of the B8 Building.

Log Id

As Print

Lahr, Germany

Photographs: Reiner Blunck

The starting point in this Project was the clients´ belief that every enterprise can and must be distinguishable from others, and should have an attractive, unique image.
The premises house both the production and administration departments of a printing firm. The two areas are connected by a glasshouse, well stocked with plants, containing the entrance and reception hall, cafeteria, lounge and conference area.
The wish to use only material that posed no danger to human health was problematic inasmuch as insulating materials such as foam glass are very expensive. New, alternative materials have not yet been sufficient tested, and so the original plan for a lightweight factory was abandoned in favour of a massive construction.
The building is extremely well insulated, and as the printing machines give off a considerable amount of heat, only minimal conventional heating is necessary in the production area.
For reasons of fire and noise protection, this section had to be shut off form the glasshouse, unlike the office block. The doors between these two areas can be left open, allowing passive heating of the office space and again minimizing the need for conventional heating.
Daylight enters through the glasshouse, filtered by the greenery, and windows – small enough to avoid glare – on the south side of the office block.
The protection of nature and the creation of a healthy working environment were important issues not only in the construction stage but also in the design of the day-to-day running of the factory; for example, only ecologically acceptable printer´ s dyes are used, and whenever possible, recycled products are the norm. Furthermore, temperature and lighting systems are regulated constantly by automatic control.

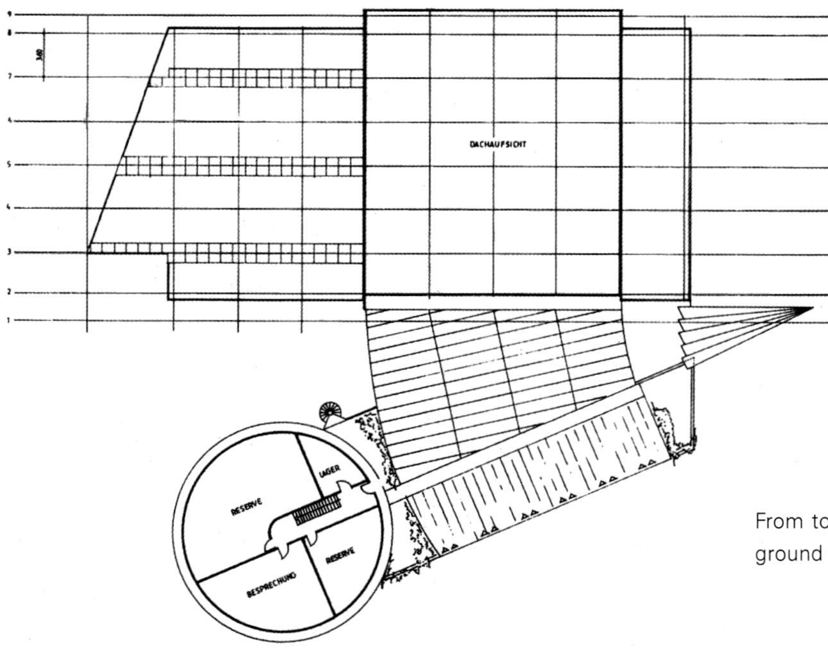

From top to bottom, plans of the second, first and ground floors.

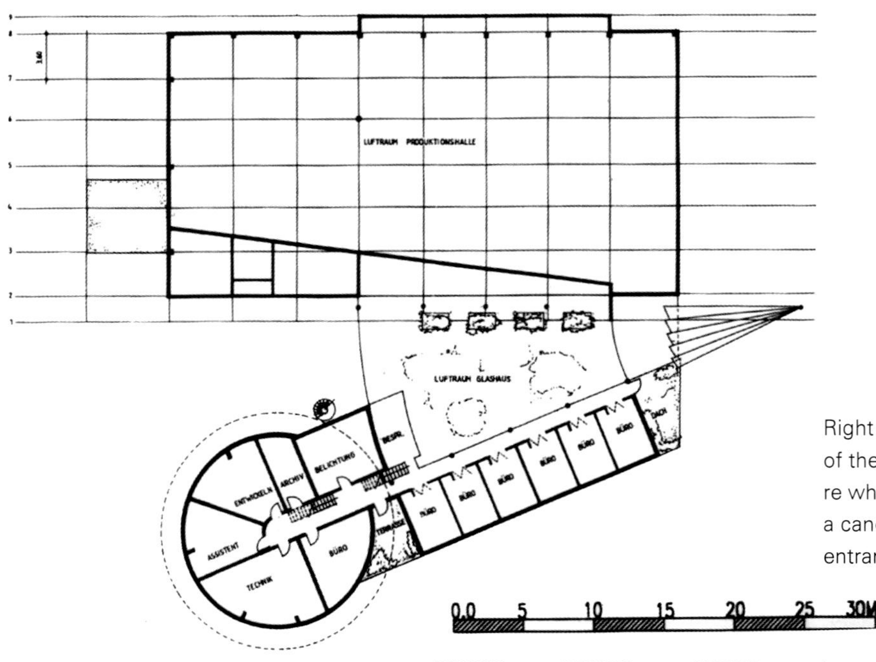

Right: the building is located in an industrial district of the town of Lahr, and is reached by a thoroughfare which runs parallel to the main road. On the east, a canopy with a striking triangular design marks the entrance to the building.

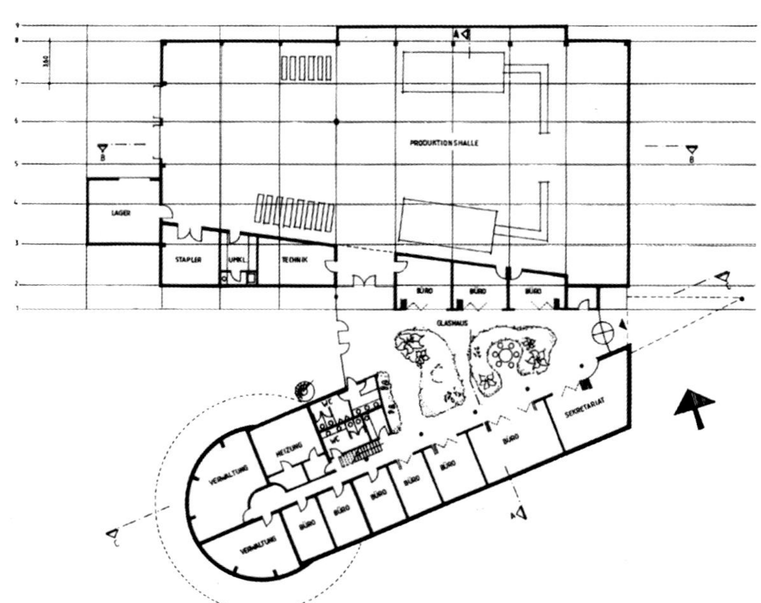

31

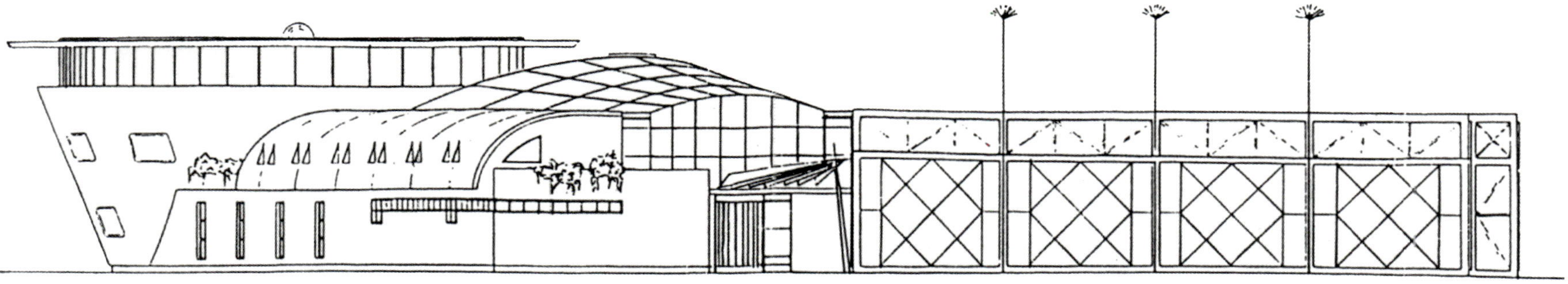

31

31

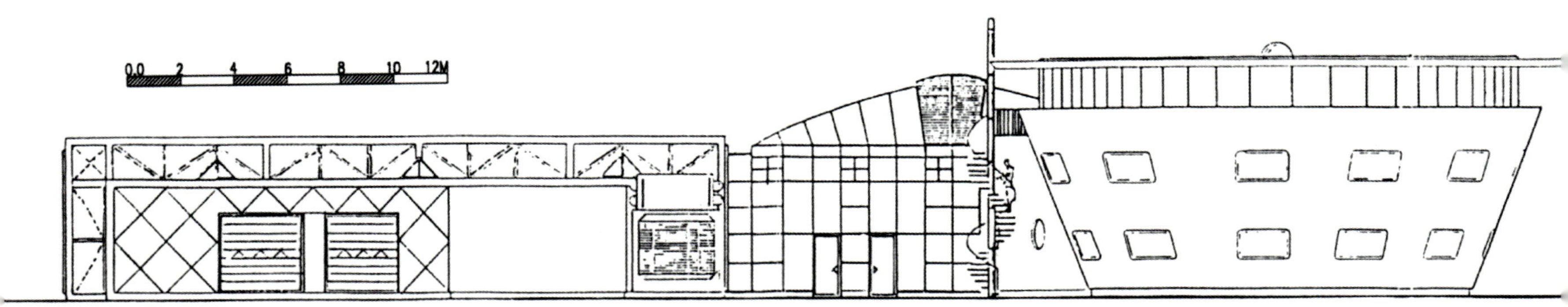
0.0 2 4 6 8 10 12M

The Project seeks total integration with its surroundings, down to the tiniest detail; for example, the outdoor paving is not cemented. Moreover, the use of ecologically acceptable materials both at the construction stage and in the activities of the firm makes for a healthy, comfortable work environment.

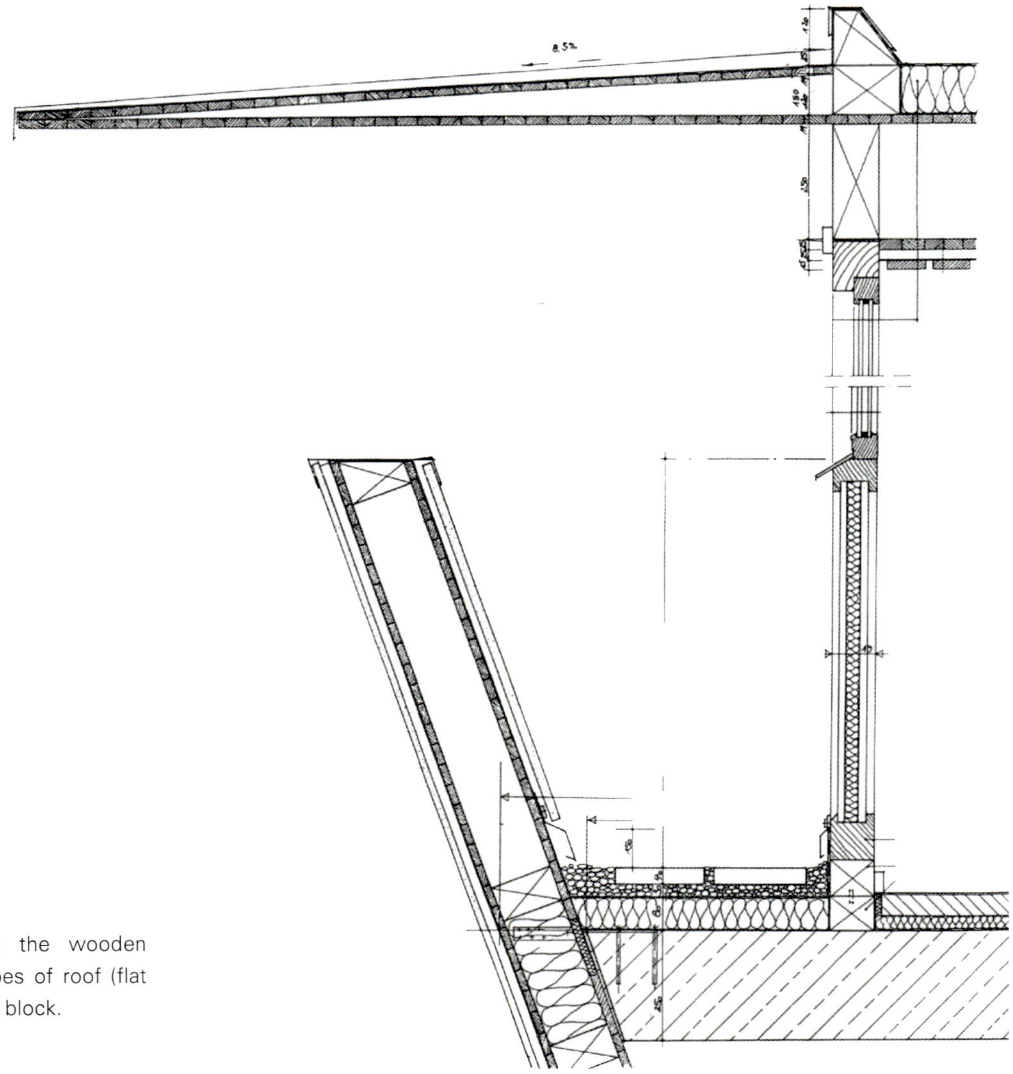

Two sections showing the wooden structure of the two types of roof (flat and domed) of the office block.

The finish of the roofs and walls offers an elegant interplay of colour and texture: the smooth plaster of the walls, the pattern of the zinc sheets of the vaulted roof and the crisscross of the truncated cone structure.

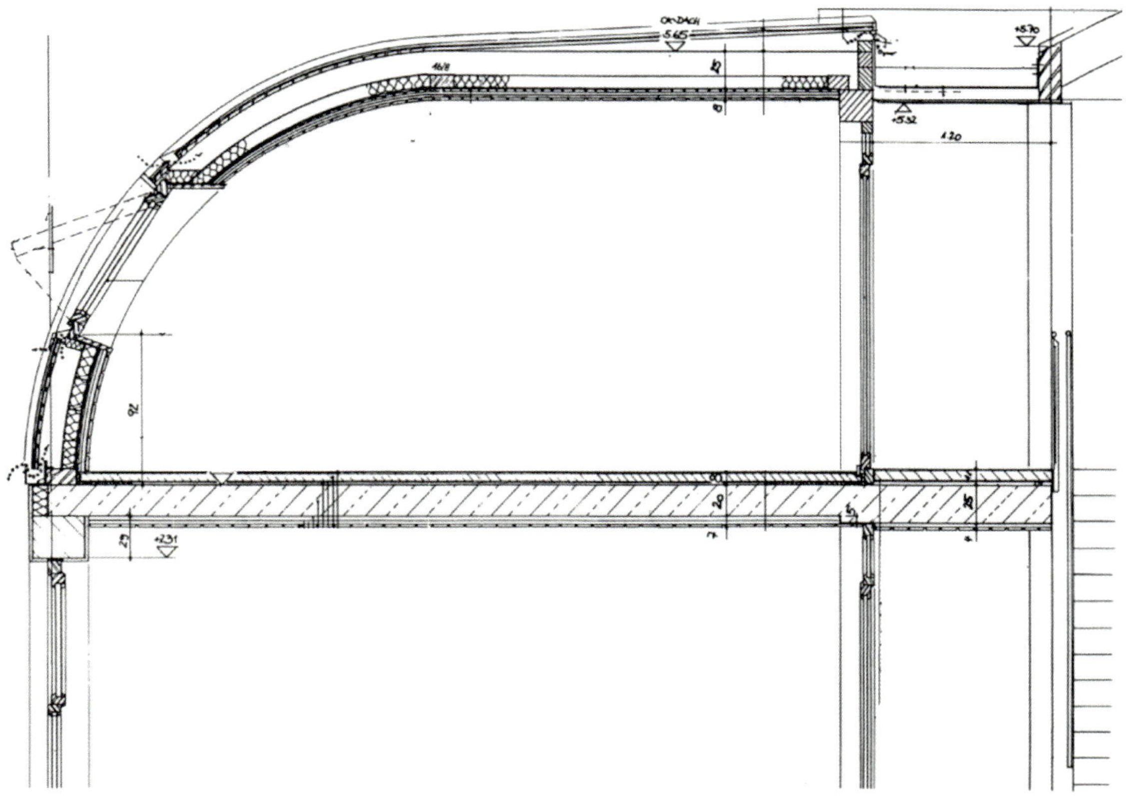

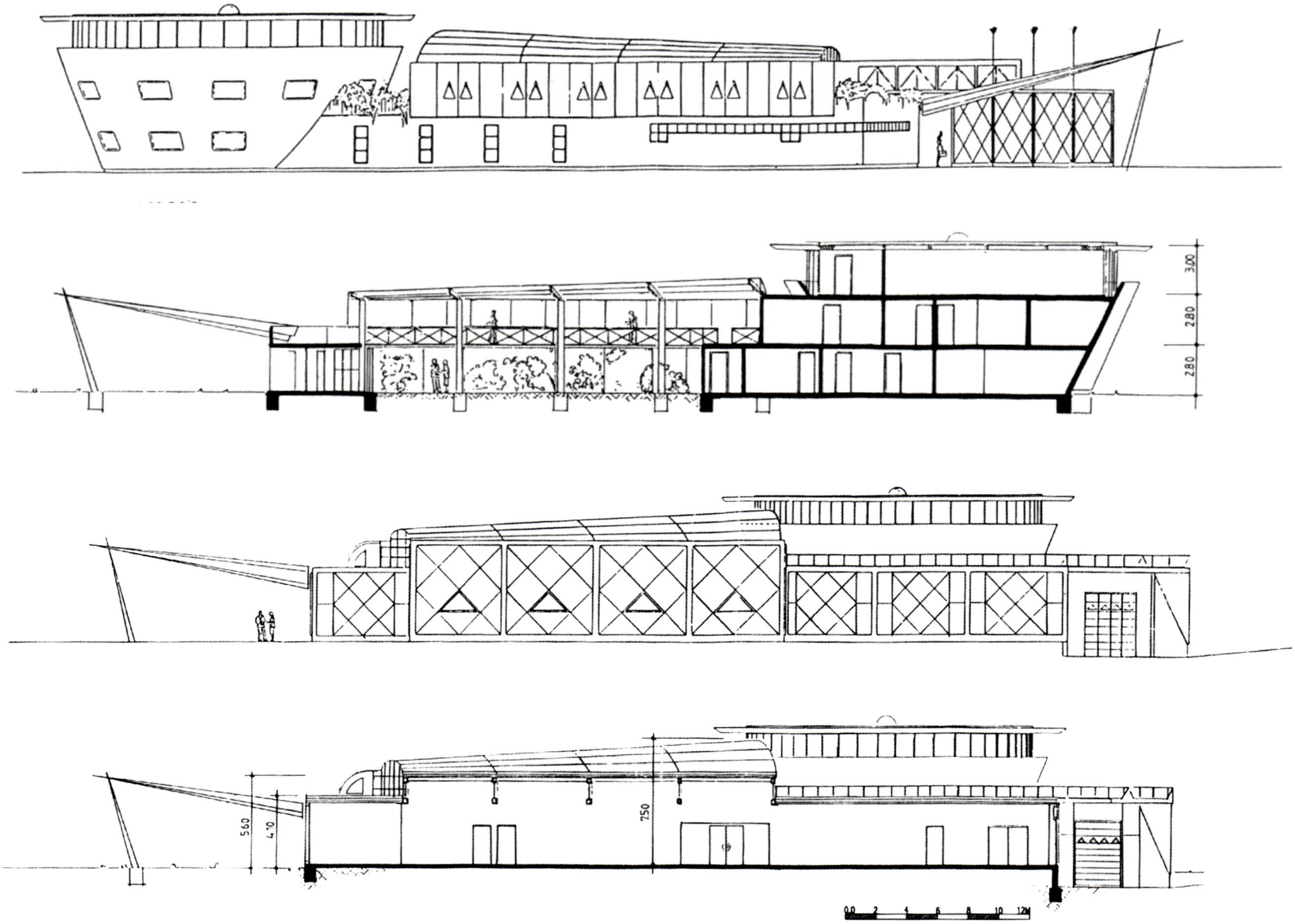

Various sections and elevations, interior views of the glasshouse and diagram f the environmentally friendly features: solar energy and utilization of the heat produced by the machinery.

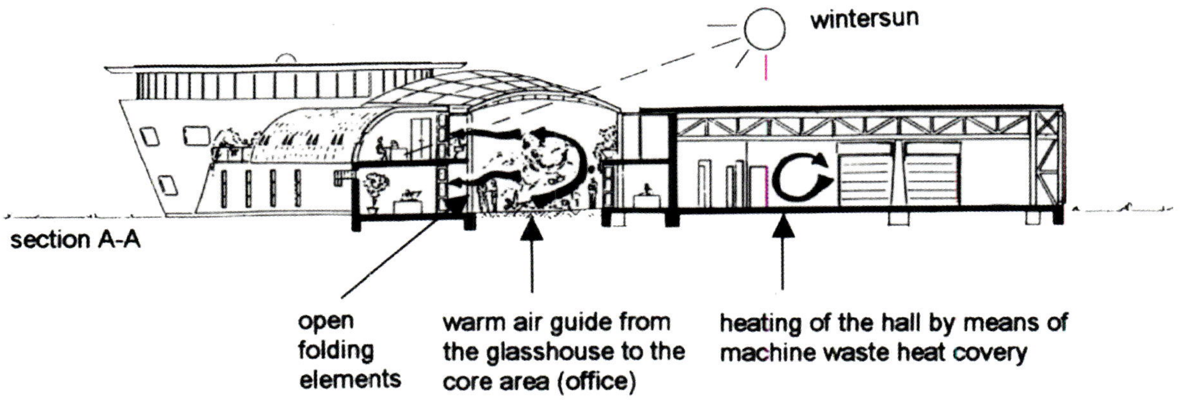
wintersun
section A-A
open
folding
elements
warm air guide from
the glasshouse to the
core area (office)
heating of the hall by means of
machine waste heat covery

Nick Derbyshire

Willesden Freight Depot

London, U.K.

Photographs: Paul Childs

Nick Derbyshire Associates Ltd is a company specializing in the design and planning of transport facilities and infrastructure, having inherited the functions and key staff of British Rail´s former in-house architecture and design group.

The 1.450-square-metre, £1.4m Willesden Customs and Public Health facility forms part of the major £8m reconstruction of Willesden freight depot for Rail Freight Distribution, a division of British Rail.

Situated alongside the West Coast Mail Line, five miles north of Euston Station in London, the design team were aware of the publicity value of such a conspicuous site from the outset. The form of the building is derived form is function, coupled with the instantly recognized from a passing train.

Its function resulted in a building with a definitive front and back. Trains are loaded and unloaded on the flat side of the building, where, protected by a canopy, there are industrial roller doors providing direct access to the ground floor working area, which is built up to the same level as the rolling stock.

The administration office areas on the first floor and have windows along the length of the building, protected from glare by a brise-soleil. This upper floor is accessed by two glazed recesses on the curved facade.

The firm won the commission competitively in February 1992 and construction was completed on schedule in June 1993.

The building is characterized by its two distinct longitudinal facades: the flat façade presents openings without discontinuity (rolling doors on the ground floor, windows and jalousie that covers the facade of the first storey); the opposite facade is curved with a rhythm of small paired windows similar to those of trains, interrupted by the straight glazed surface of the stairways.

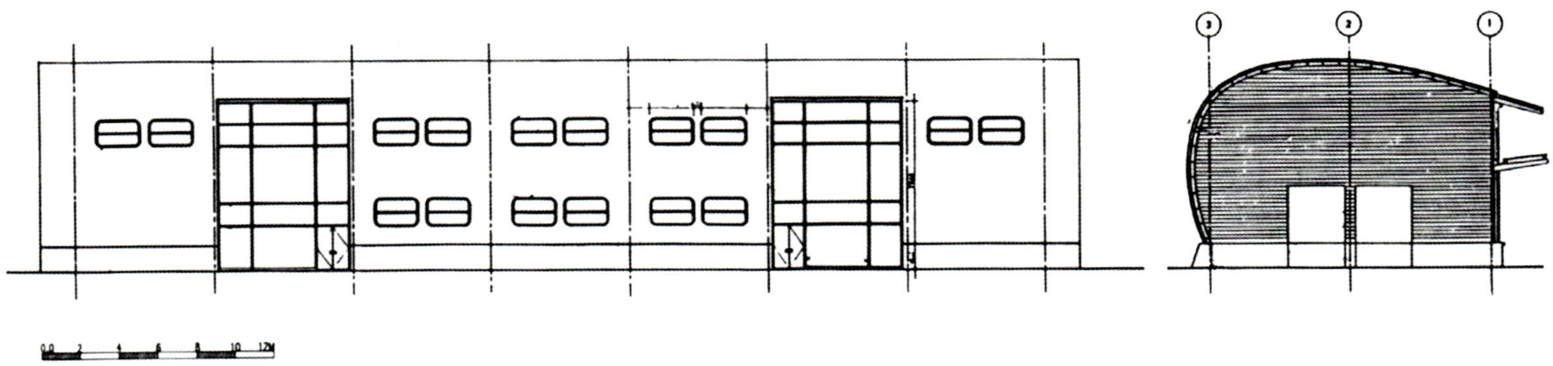

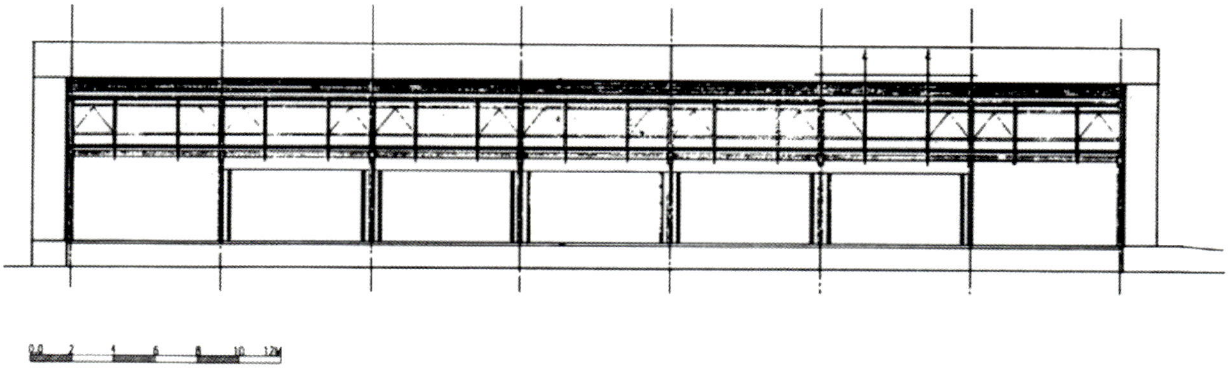

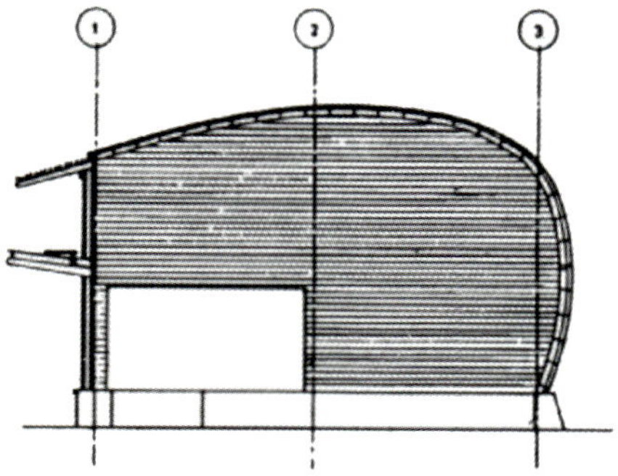

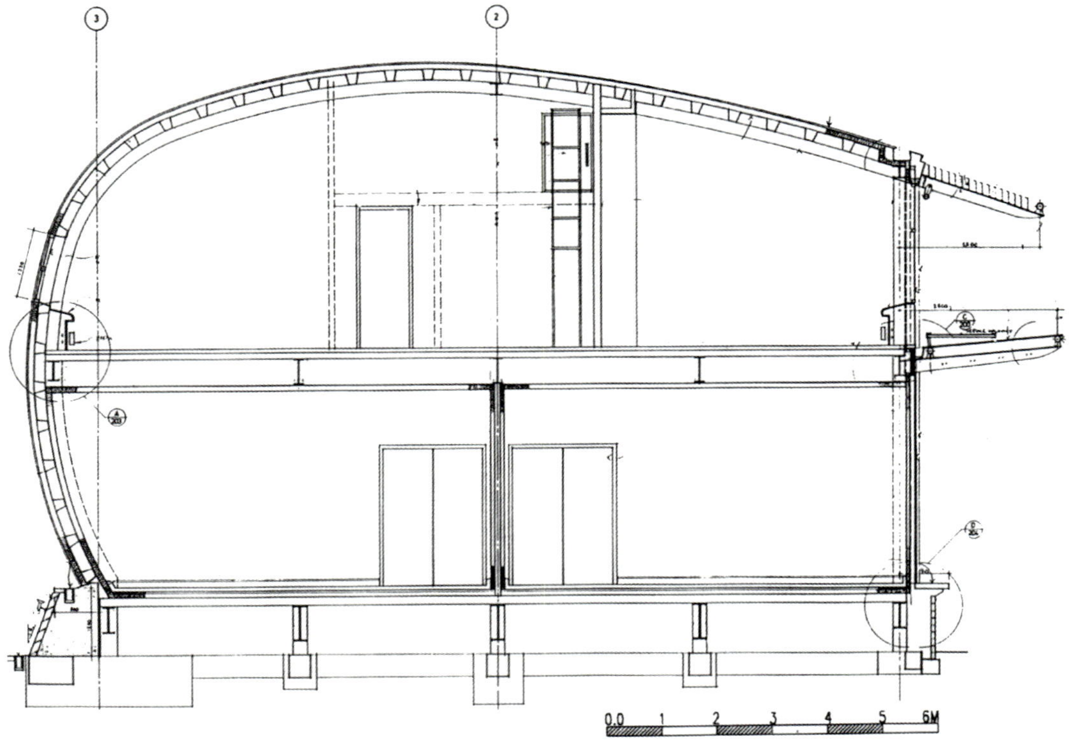

To avoid constant noise interference from the railway traffic, the offices are located on the first floor, protected by a brise soleil that runs along the uninterrupted row of windows.

Ground floor

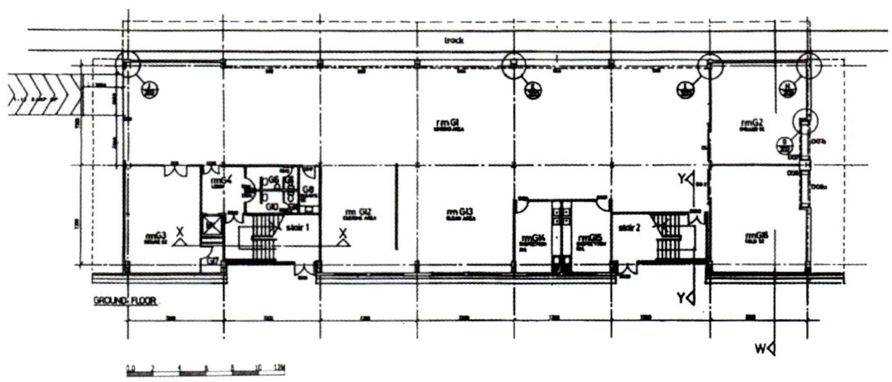

First floor

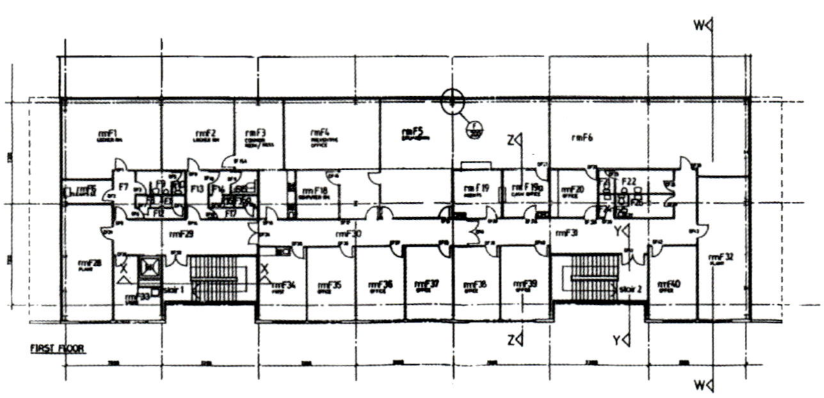

Aluminium is the material that dominates the construction in the interior (contributing lightness to the stairway) as well as the exterior, where its grooves offer an elegant vibration.

Sergio Maria Calatroni

N.O.Z office building

Shizouka, Japan

Photographs: Sergio Calatroni

The N.O.Z. office building in Shizouka is, together with the Kashiyama shop in Paris and the Fujitaka restaurant in Milan, one of the most important projects performed by Sergio Maria Calatroni.
One of the aims was that the building should be an emblematic representation of the company: the result is excellent. The building is developed through a startling interplay of volumes, perhaps the result of the close relation that the architect has with sculpture, giving each space a different form and each floor an independent distribution. Calatroni pays special attention to the way in which the light penetrates the interior, since it is what explains and qualifies the spaces. His building thus becomes an explosion of light and colour.
The plain and coloured opaque surfaces of the interior are set against the polished and practically immaterial surfaces of the exterior. The units of metal plate adapt to the curvilinear forms of the facade, reflecting the sunlight.

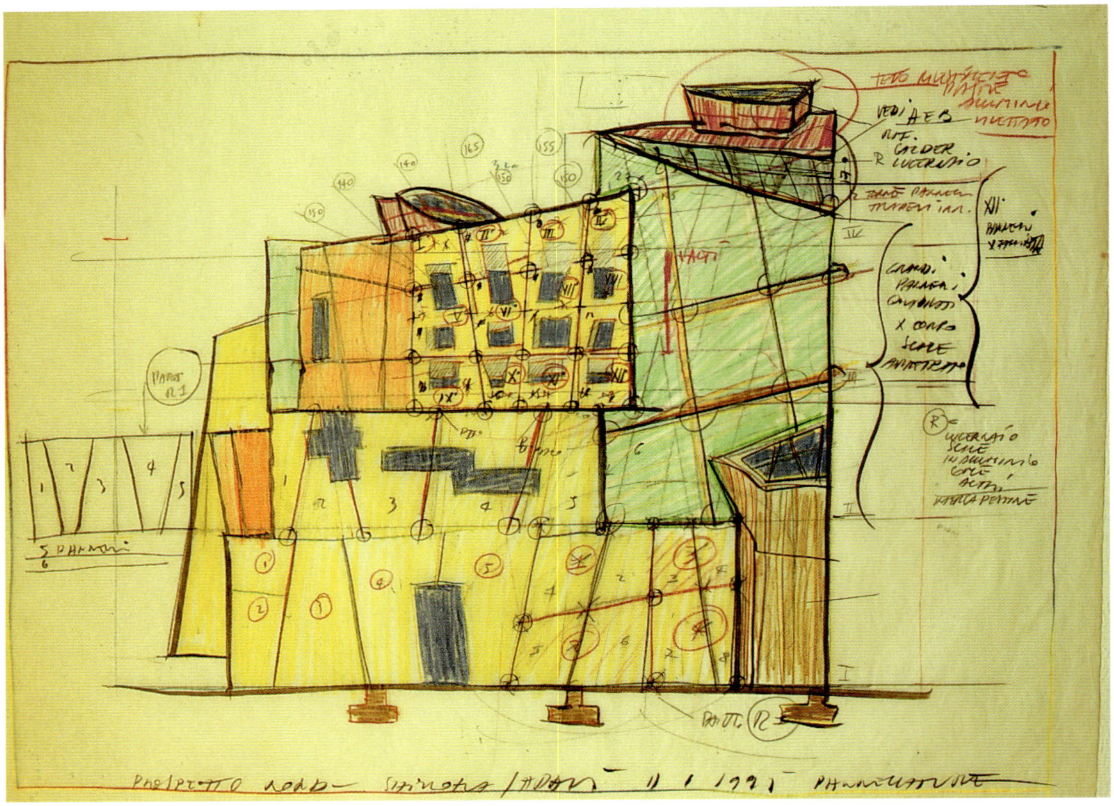

CA
杉山たば
TEL. 45-
DyDo

The building is Developer through a startling interplay of volumes. The metal plate units adapt to the curvilinear forms and volumes of the building.

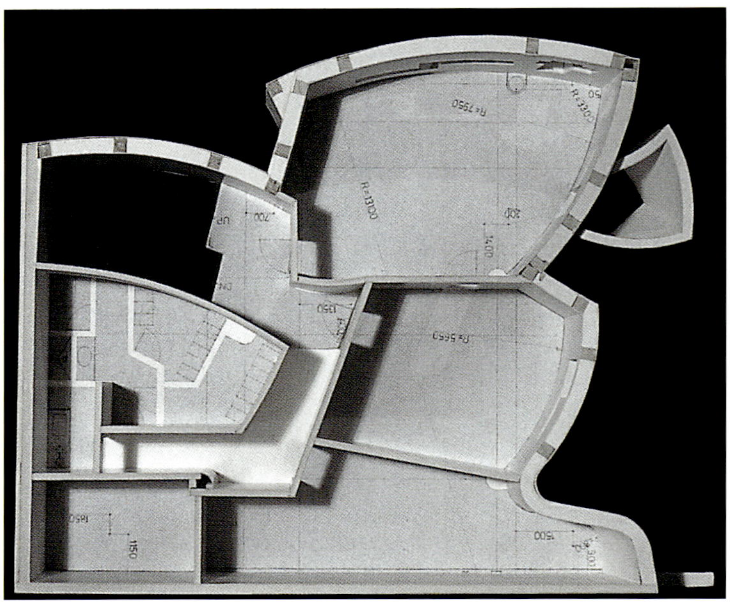

The light qualifies and defines the spaces

Detail of the skylight covering one of the halls

Nicholas Grimshaw

RAC Regional Control Center

Bradley Stoke, Bristol

Photographs: Peter Cook /Hilary Coe

The new RAC Motoring Services building forms the newest part of the strategy to provide high-profile, technologically advanced centres adjacent to major motorway interchanges.

The car park has been set out axially on the entrance to the building. From the raised level of the car park mound, visitors and staff enter the building on a bridge to arrive at first floor level, the middle of three floors of accommodation.

The building is set into a landscape bowl improving the immediate surroundings and sheltering the ground level from the noise of the adjacent motorway.

The building is almost totally glazed with the necessary noise reduction resulting form the sealed external glazing. The outward tilting external wall is shaded with external walkways designed to provide both maintenance access and act as sunshading louvers. The silver-grey roof overhangs allow maximum views and daylight without glare. The external wall transmits the loads imposed on the glass and the external walkways to the inner spars of s taut steel structure. The colours of the building are deliberately neutral so that the building can provide a transparent framework for the human activity within and the corporate activities presented to the outside world by the RAC.

The plan of the building provides the maximum of efficiency in organisational use as well as maximum daylighting and views for staff and visitors. The office accommodation on three floors wraps round a full-height central atrium. An inner ring of glazed meeting rooms enjoys good views into the atrium and additional daylighting form roof glazing. An open staircase within the atrium helps create a sociable working environment with stairs and light visibly connecting all levels.

On the outside, the building becomes a transparent skeleton that shows the activities that take place in the interior.

South West Elevation

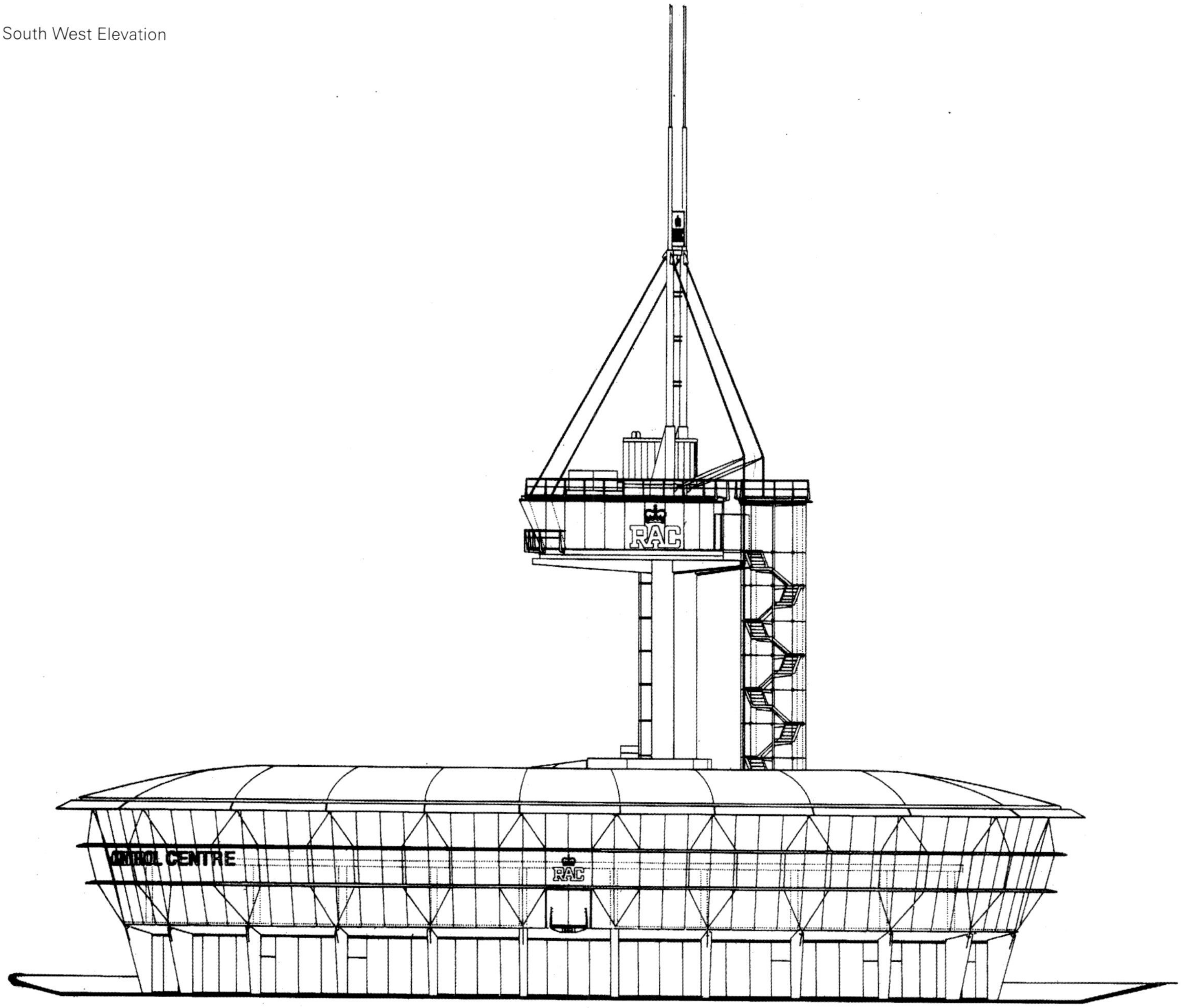

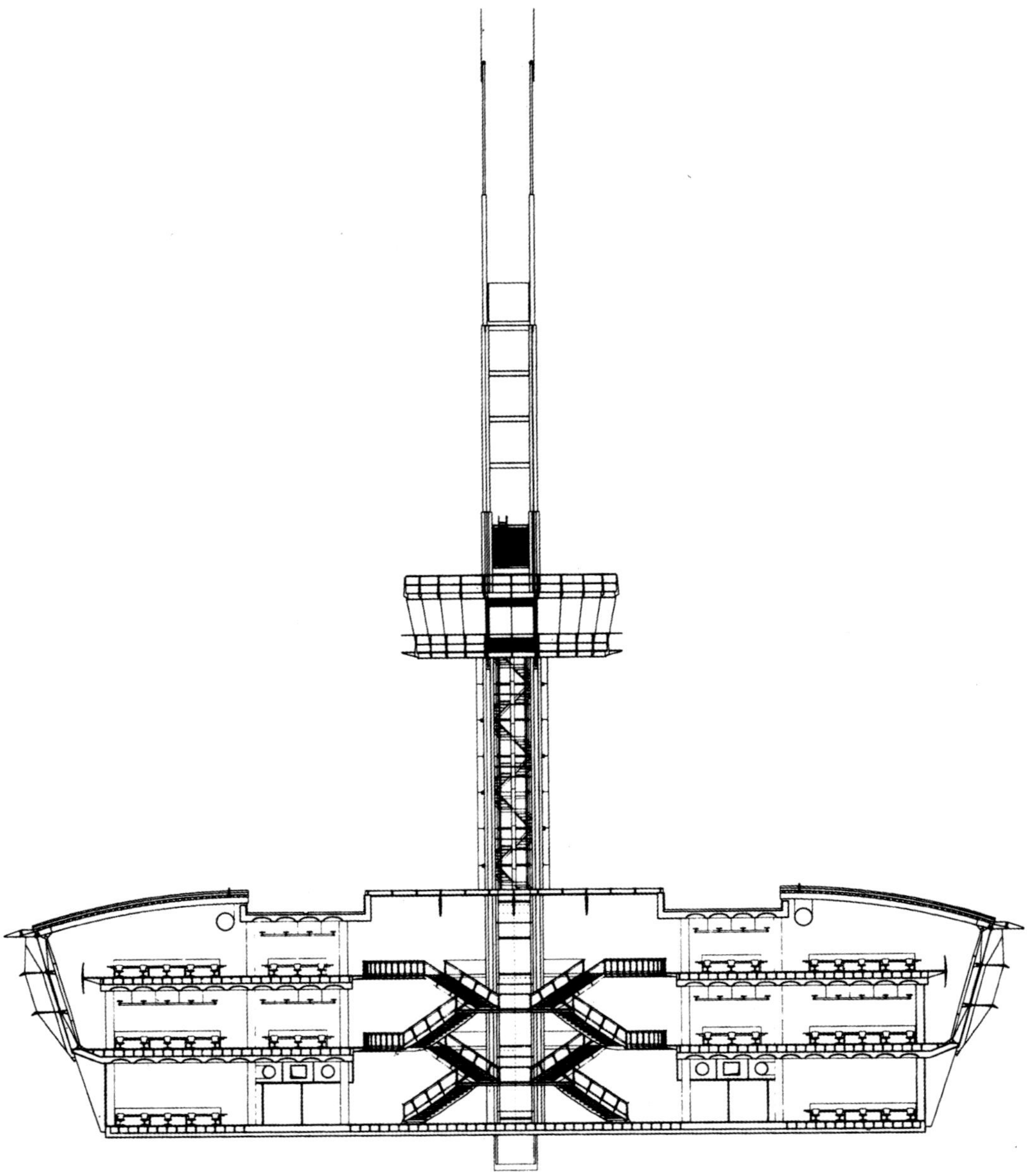

Cross section

Ground floor plan

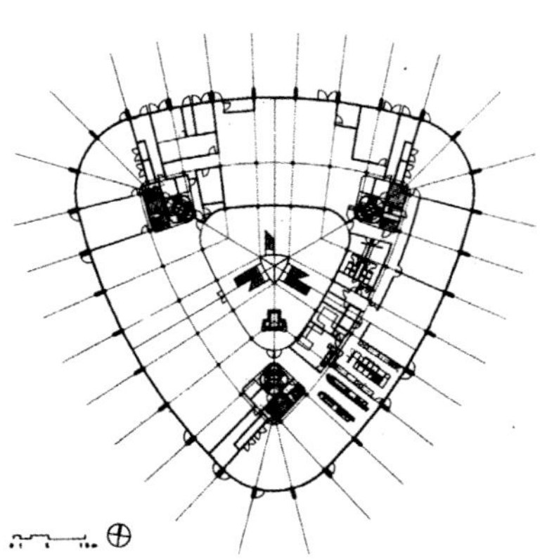

First floor plan

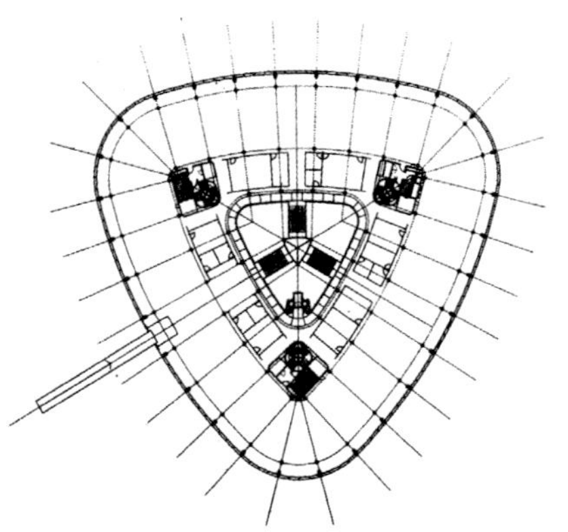

Upper floor plan

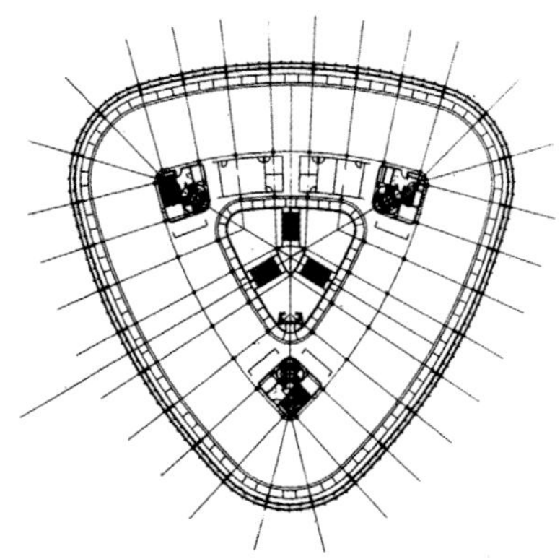

Post Room

A belvedere situated 35 metres above ground level provides spectacular views of the surroundings.

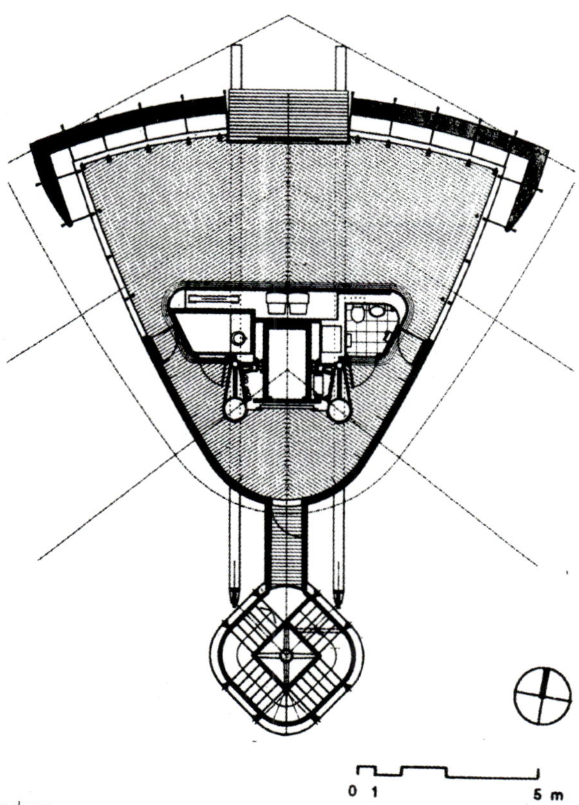

Plan of the belvedere

The building, which is practically totally glazed, has external passages that give access for maintenance and visors that protect it from the sun.

Neutelings Riedijk Architecten

Veenman Printers

Ede, The Netherlands

Photographs: Christian Richters

The new building for Veenman printers in Ede consists of two sections: the printing works with the necessary stock room and an office section.
The site is in a very typical business location amongst a lot of fairly randomly dispersed buildings of mediocre quality in which the office section is always differently articulated and materialised than the production section.
In order to avoid this traditional distinction between white and blue-collar workers, the architects opted for a continuous elevation wrapped, as it were, around the building.
In addition, the building is organised around a courtyard garden from which the various departments are in visual contact with one another. This courtyard garden (designed by West 8) forms a counterpoint to the neighbouring business premises and creates a pleasant work climate that is in touch with the changing seasons.

The facade is composed of three sections: a black, concrete stone plinth, a row of windows in the plinth and above this a glazed section. The latter consists of glass panels trough which it is possible to see the light-reflective insulation. The glazing is mounted in a simple framework derived from greenhouse construction. The glass is printed with a text by K. Schippers (layout by K. Martens).
The building has a V-shaped roof. The section is such that the areas requiring a lot of head-space are located under the highest sections of the roof. The volume is indented at two spots: the main entrance, which is marked by the load-bearing company logo, and the dispatch entrance. These, together with the V-shaped cornice line, add up to a striking volume that can be recognised from the motorway.

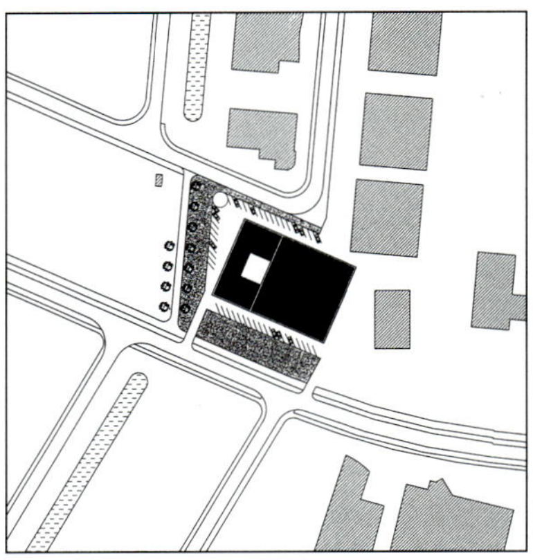

The building has an imposing presence in the landscape. The shape of the roof breaks the characteristic monotony of industrial buildings, whereas the facade, with its letters, evokes the nature of the company. A text of K. Schippers, designed by K. Martens, is etched in the glass.

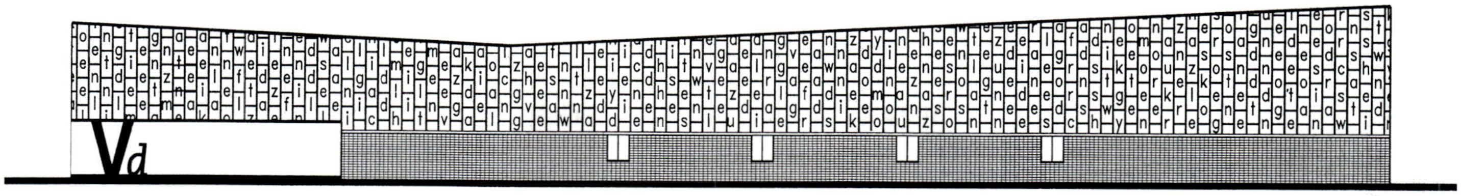

South elevation

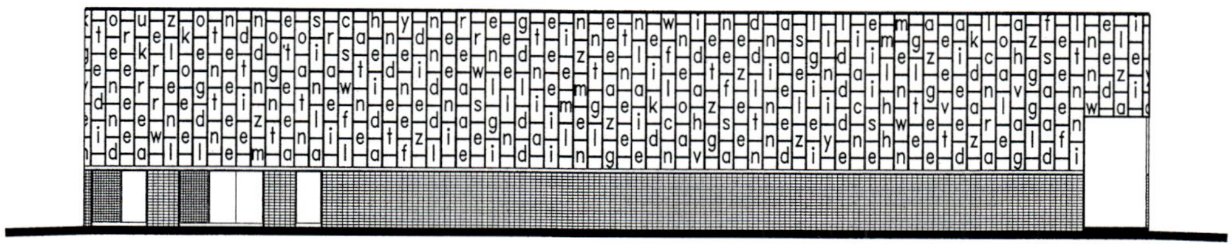

East elevation

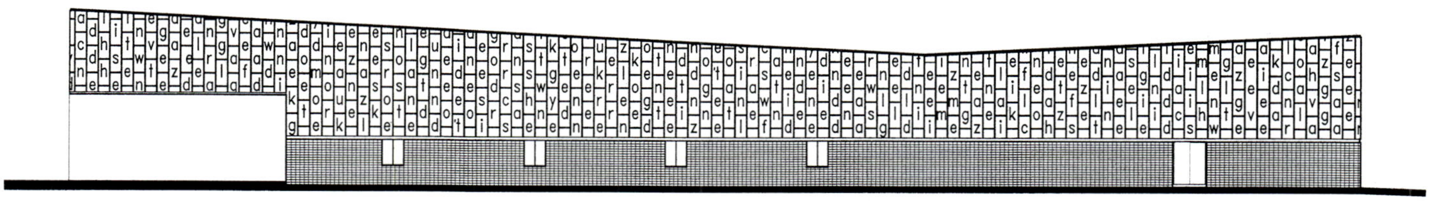

North elevation

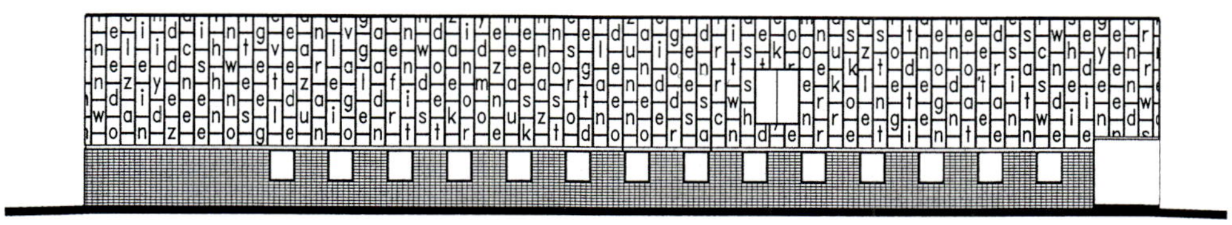

West elevation

The facade of the building acts as a screen, concealing its true structure. On the following page, views of the inner court, around which the different depart- ments of the firm are articulated. The outer cladding of wood gives the space a warmth that is not common in buildings for industrial use.

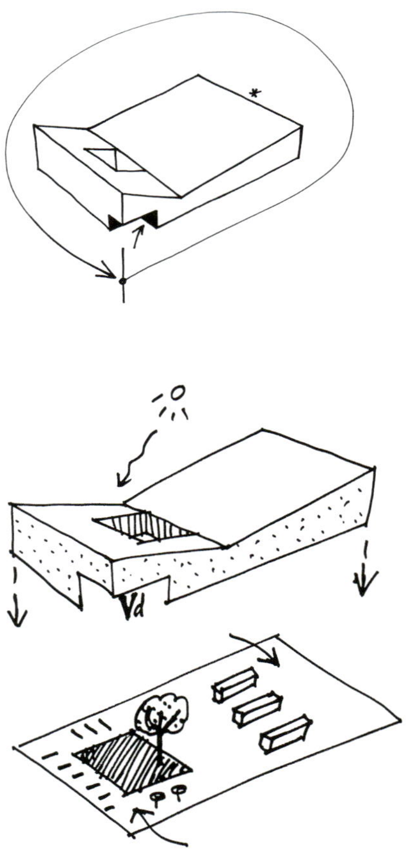

Scketches

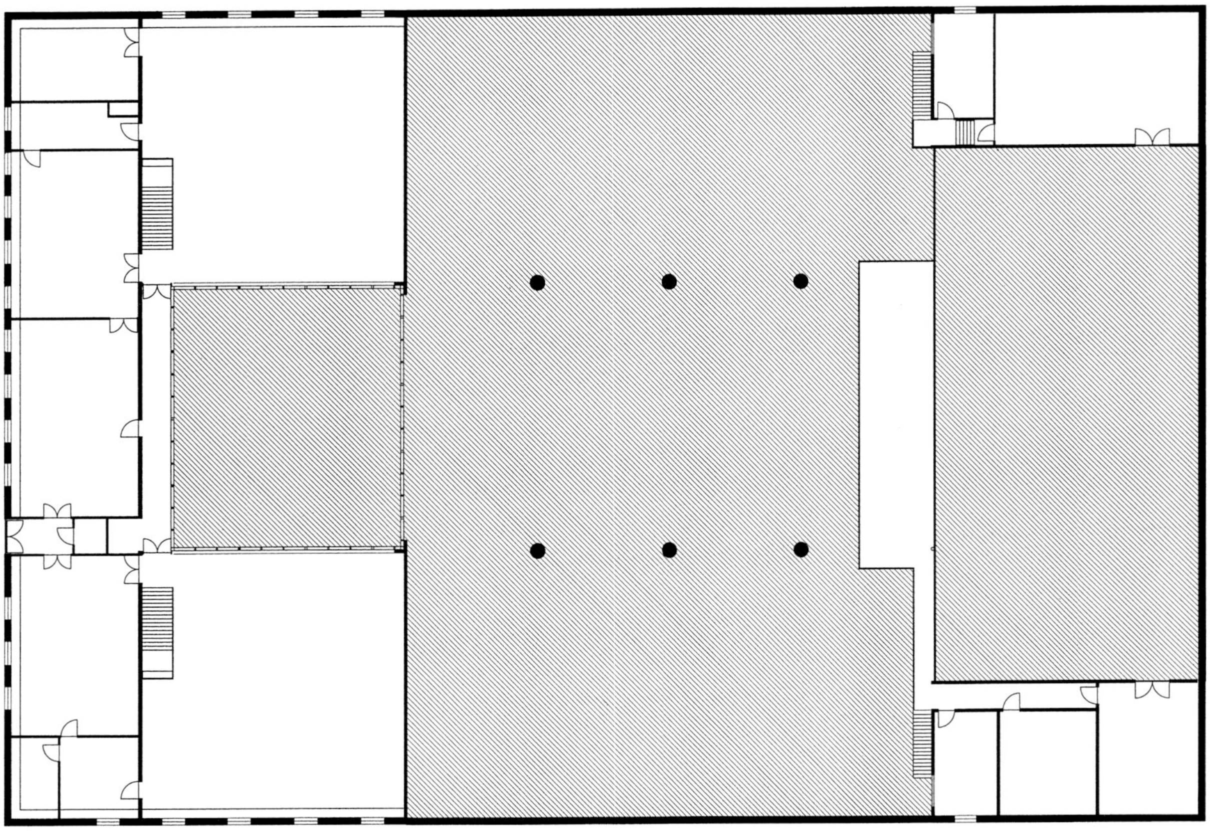

Upper floor

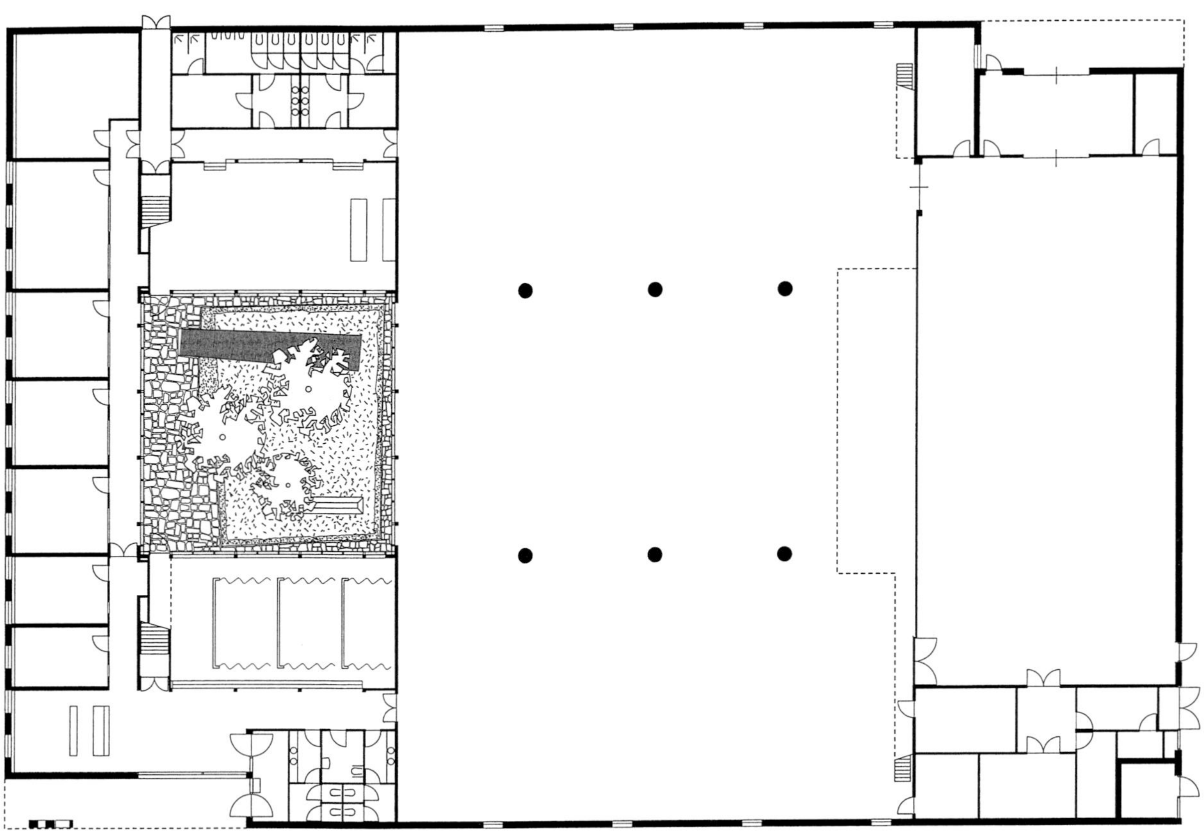

Ground floor

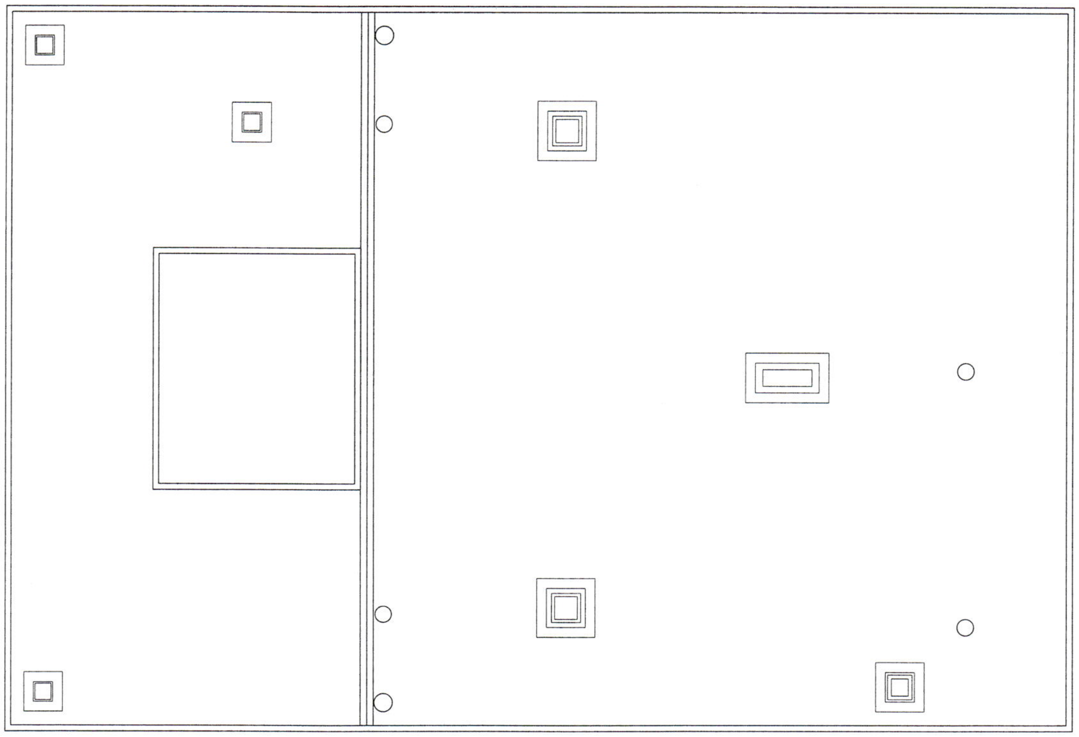

Roof level

0 2 4 6 8 10

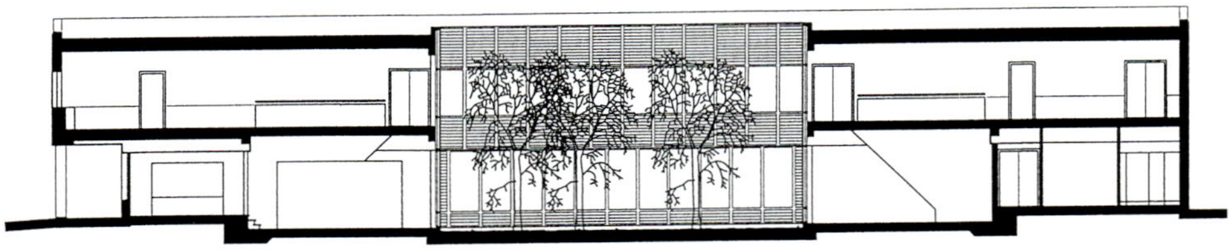

Cross section through internal court & production area

Longitudinal section through internal court & production area

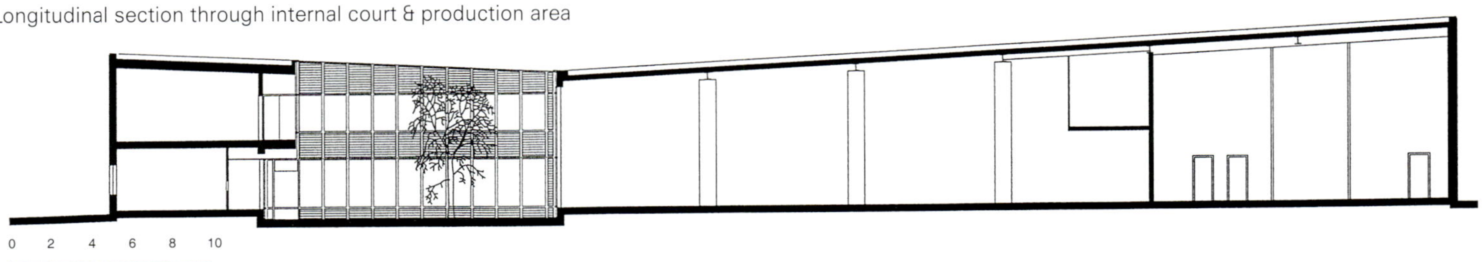

Simon Conder Associates

Georgina von Etzdorf Workshop

Salisbury, UK

Photographs: Simon Archer + Chris Gascoigne

In the early nineties, Georgina von Etzdorf, the textile design and production company, decided to group all their quality control and administrative functions in one place. The site they chose for their new building accommodated a farm shed but was surrounded by trees with outstanding views to the south and west over a valley. However, the local planners and politicians insisted on a scheme that could be presented as the conversion of a redundant farm building, even if in reality this has meant only retaining the main portal frames from the original building.

The clients brief was to design a building that would allow them to consolidate all their quality control and administrative functions on one site and would do so in a way that would reflect the open, relaxed and non-hierarchical culture of the company and fully exploit the landscape potential of the site. The solution was also to be capable of ready internal adaptation and extension to meet changing and expanding needs. These objectives were constrained by a very tight construction budget of £35 per square feet.

The scheme consists of the original portal frames extended to the west and clad in black corrugated steel, with large areas of glazing on the southern and western sides of the building to visually link the internal working environment to the surrounding landscape. In the future additional bays are planned at the western end of the building. Internally the building is essentially one space with the storage areas on the north side separated from the quality control functions on the south side by a large plywood structure. At ground level this structure contains the sanitary accommodation, storage, a meeting room, a kitchen/dining area and the reception area.

Materials and garments can pass through this structure at two points, from goods-in to quality control and then back to goods-out. At first floor level there is an open gallery for the various administrative functions, naturally lit by a long northlight and large windows on the east and west facing gable end walls. The plywood sheets that form the walls, ceilings and floors of this structure are screwed to the softwood studs and floor joists to facilitate layout changes in the future. At night the interior is lit by a combination of metal halide uplighters fixed to the plywood structure at first floor level and similar mobile uplighters located adjacent to the mobile workstations at ground level.

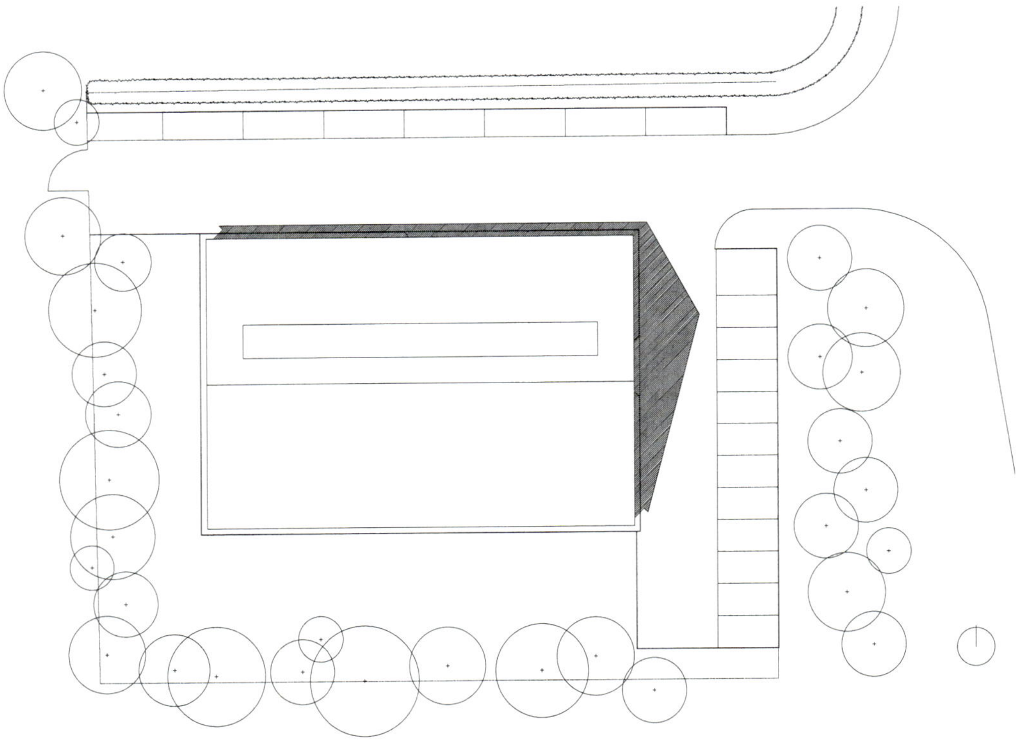

Site plan

The east elevation, which is used for goods-in. The west elevation represents the end of the quality control process (goods out). The asymmetric arrangement of the windows provides dynamism and tension.

Black corrugated steel panels were used for the walls and roof. This material underlines the industrial character of the building and contrasts with the surrounding greenery.

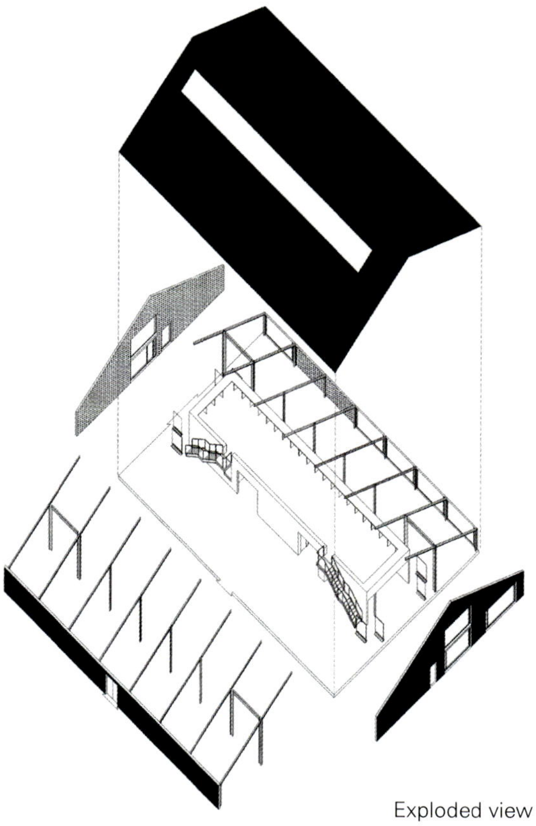

Exploded view

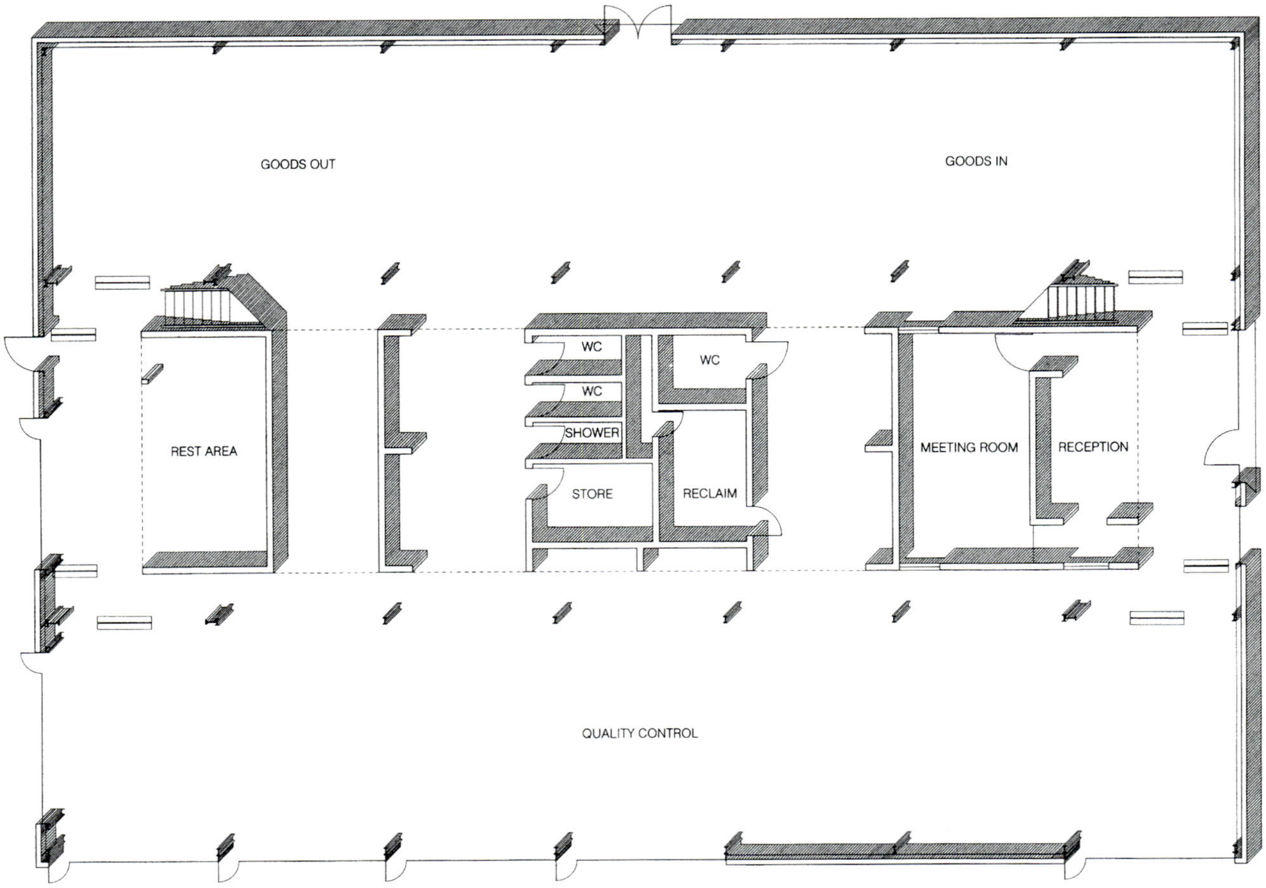

Ground floor plan

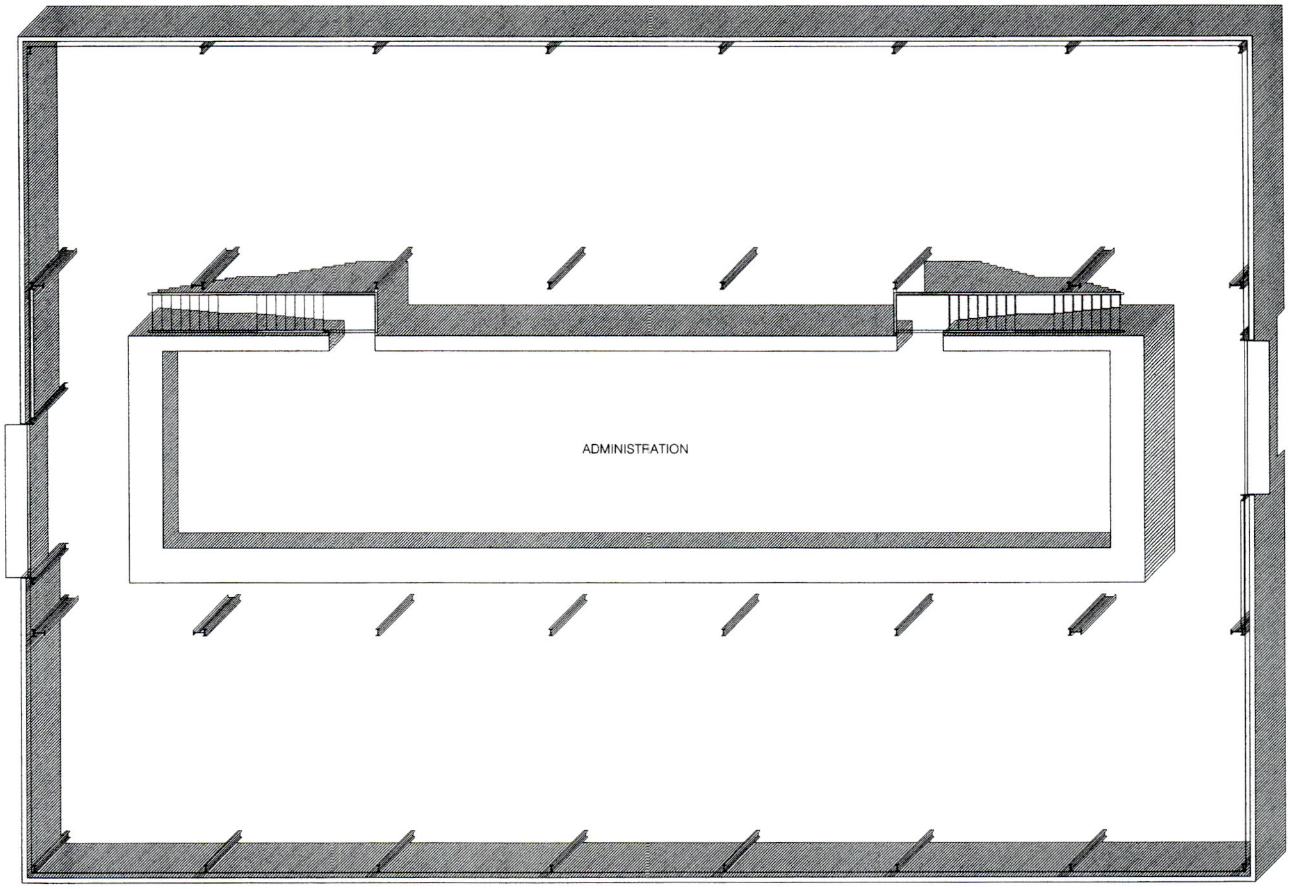

First floor plan

Barkow Leibinger Architekten

Addition to Haas Laser Factory

Schramberg, Germany

Photographs: Margherita Spiluttini

Located on two hectare site on the periphery of Schramberg in southern Germany, this project doubles in size an existing facility of 3,000 square meters built in the 1980s. The new laser production facility for 60 technicians includes offices, laboratories, training spaces, and a employee café. The site, bounded by major streets to the south and north, slopes to a bermed grass divider to the highway at the north. Additional shops and administration buildings built in the 1950s share the site with the new addition.

The addition encloses an existing courtyard with the provision of a new terrace lined with a café and break rooms. The off-set and interlocking roof system re-interprets the existing roof by providing north-facing skylights that also contain the extensive mechanical equipment required for the laboratories. Construction is a concrete slab and column system with a 2.50 meter steel truss roof in a off-set layout, which allows daylight to penetrate the lab and office spaces below.

The V-shaped volume at the top of the skylights provides space for the mechanical equipment ad acts as a large lantern for artificial lighting elements.

The lower level provides temporary training facilities, labs and the truck loading and shipping area. The facades of the upper level is a series of undulating folded aluminum panels. The lower level is a continuous band punctuated by openings of clear glass mixed with translucent u-profiled poured glass sections. His arrangement produces the effect of the closed upper level floating above the light glass base.

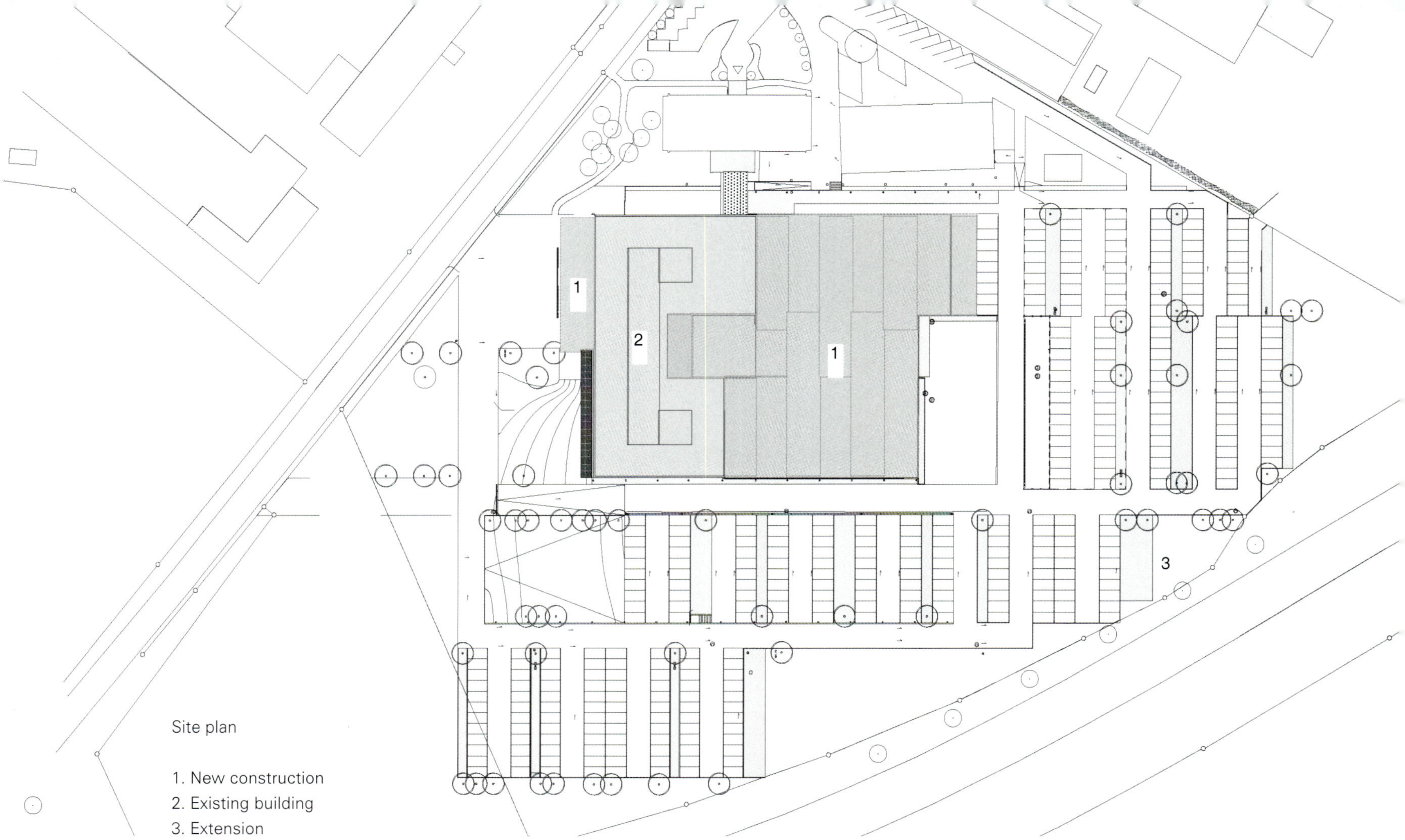

Site plan

1. New construction
2. Existing building
3. Extension

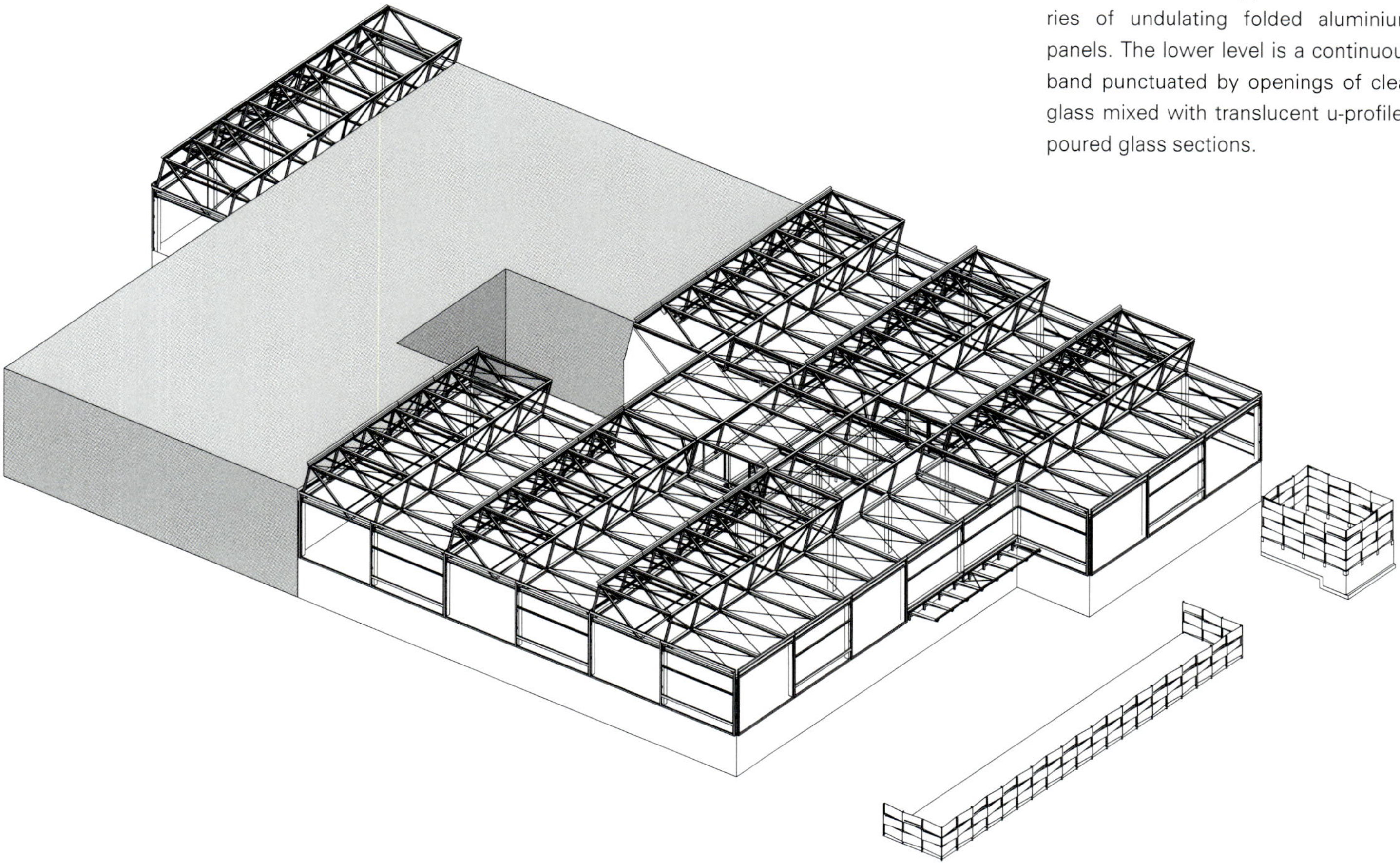

The facade of the upper level is a series of undulating folded aluminium panels. The lower level is a continuous band punctuated by openings of clear glass mixed with translucent u-profiled poured glass sections.

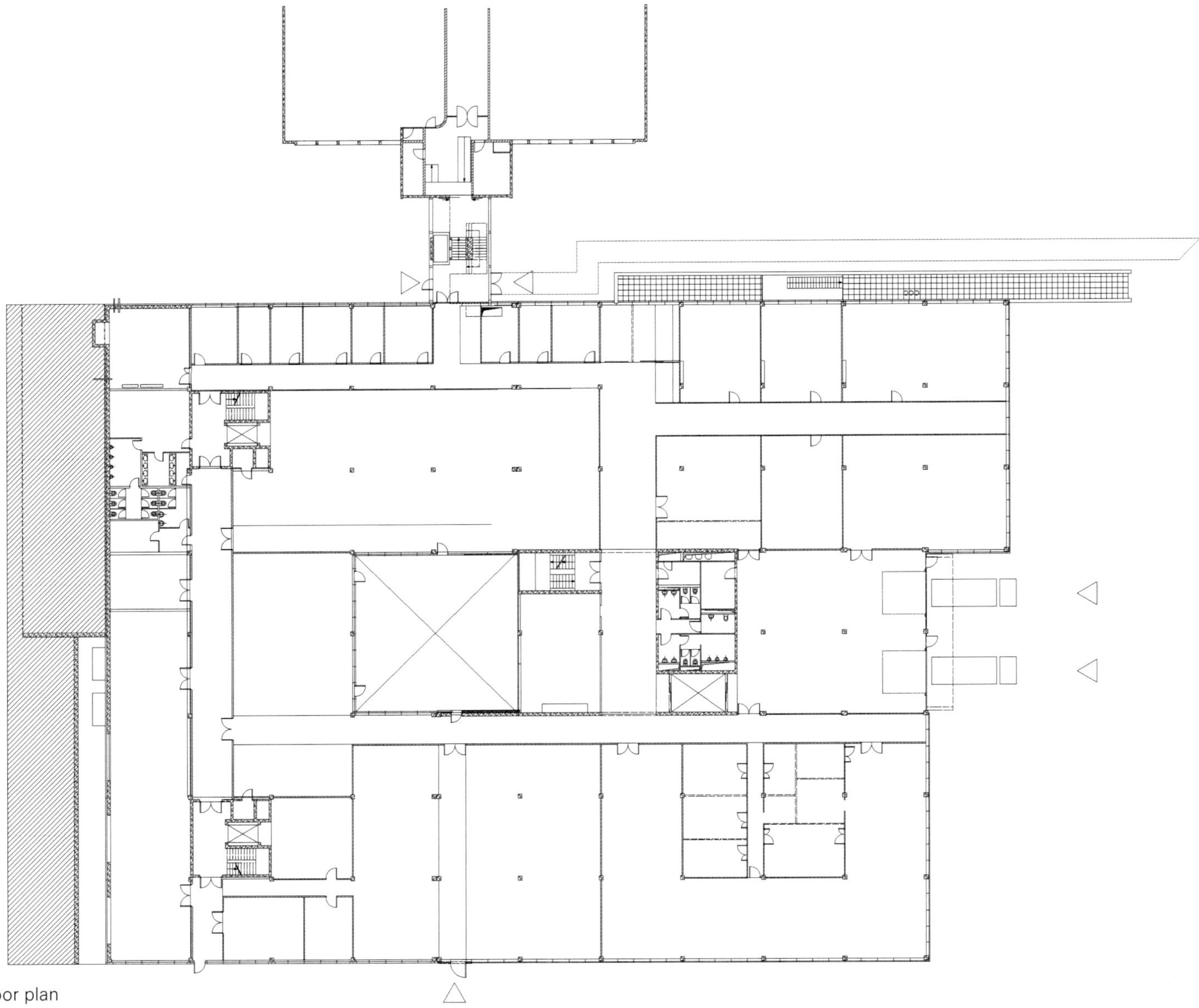

Ground floor plan

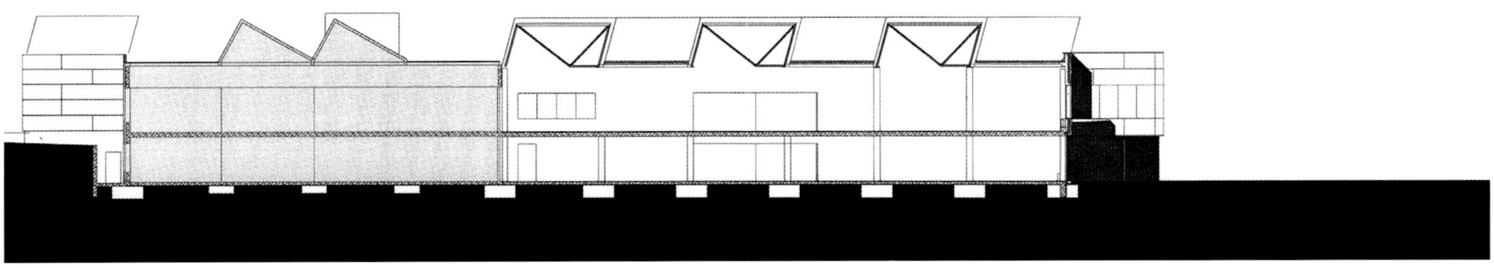

Longitudinal section

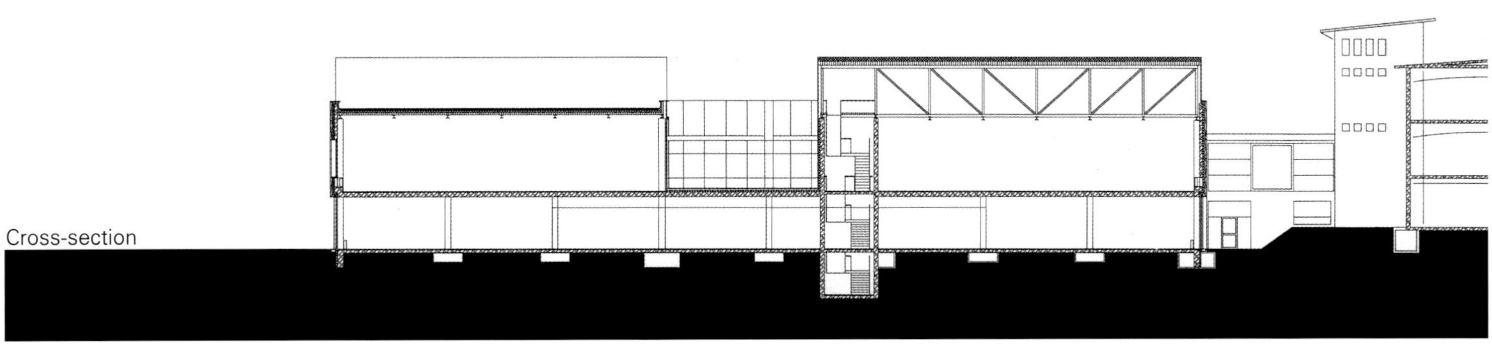

Cross-section

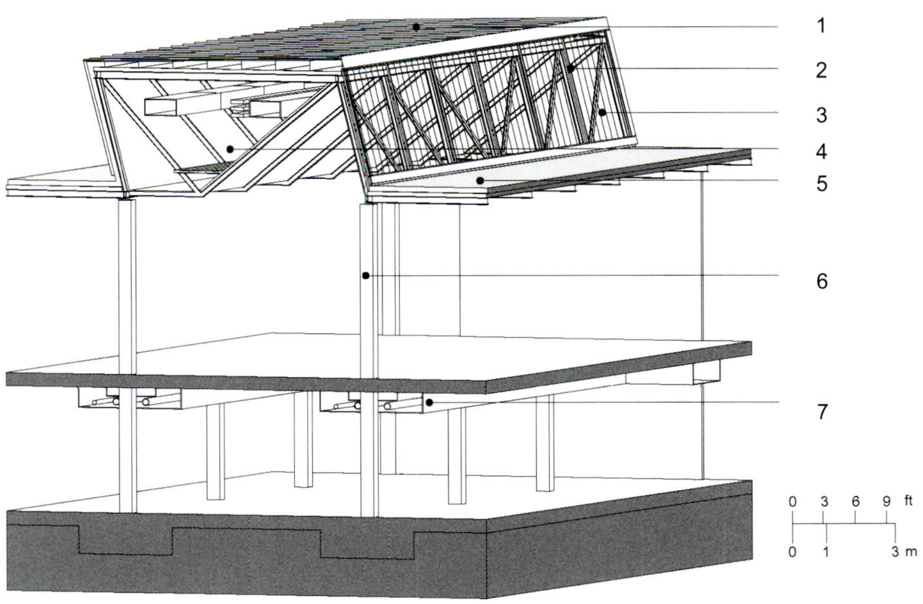

Axonometric construction

1. Thermally insulated aluminum roofing
2. Steel truss roof construction
3. Double-shell profilit and aluminium clerestory
4. HVAC/Mechanical loft
5. Gravel bed roof
6. 40x40 cm concrete column
7. Electro-mechanical service conduit

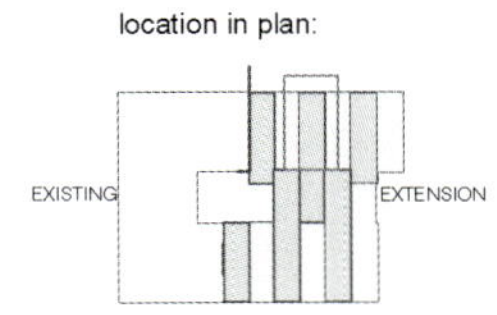

The structure is a concrete slab and column system with a 2.5 meter steel truss roof in an off-set layout, which allows daylight to penetrate the lab and office spaces below.
The V-shaped volume at the top of the skylights provides space for the mechanical equipment.

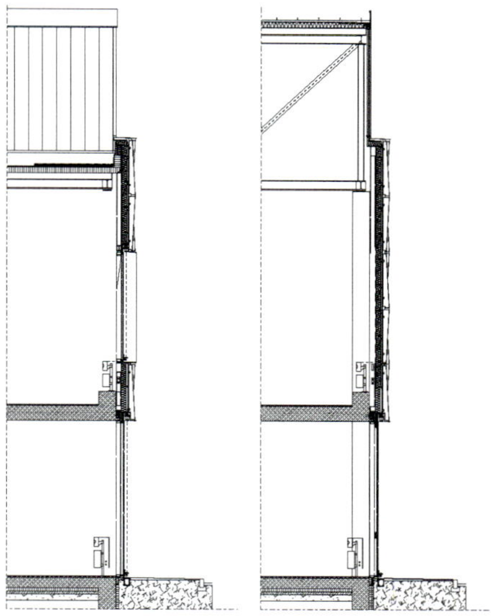

Vertical section through exterior facade

Carlos Ferrater & Joan Guibernau

Nueva sede de J.C Decaux

Madrid, Spain

Photographs: Alejo Bagué

The new headquarters of the international company J.C. Decaux occupies the old Martini & Rossi factory, a building of the early 1960s designed by the architect Jaime Ferrater and built in an industrial zone on the outskirts of Madrid.

The complex, protected by the Architectural Heritage Law, is formed by a great vaulted buuilding with a structure of arches, a volume for the administration and a series of warehouses and service spaces–the statement of a truly contemporary poetics of great formal elegance. It is a prime example of Spanish architecture of the time, characterised by a business isolated from the international circuits by the protectionism of the Franco regime. It shows, however, its peripheral nature and the desire to continue the discourse and the poetics of the Modern Movement in the purest sense, sheltered from the debate that was emerging in the rest of Europe.

The scheme deals with a current subject: the recovery of contemporary architecture with an attitude of respect for the past. It was thus decided to perform a series of non-invasive actions in order to adapt the building to the new requirements. The building was in a poor state of repair because it had been abandoned, and a series of works had deformed the original structure, both in the interior and on the facades. Due to the need to convert it for a new use, and to adapt the structure, the facilities and the communications and evacuation system to the current regulations, the scheme focused mainly on three criteria:

- Recovering the façade whilst respecting the original language, whose image had been deformed. Replacing the course aluminium frames with special new ones that meet the current regulations and restore the original appearance of the glass wall.

-The new use as both an administrative office and an exhibition hall, with the need for a series of spaces open to the public, involved the creation of a new, more representative central access located at the point of encounter of the existing volume, emphasised with a new overhanging canopy.

-The internal structure is fully respected, except for some modifications of the interior layout and a series of openings in the floor slab to improve the natural lighting in the zone open to the public.

The upper level houses the administration and management, connected on the second floor with the upper level of the showroom, an overhanging structure that forms a light metal walkway that opens onto the full-height space of the foyer.

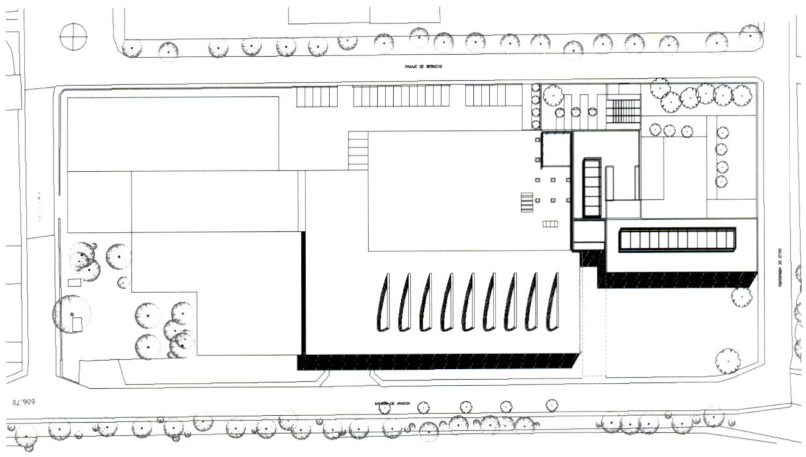

General floor plan

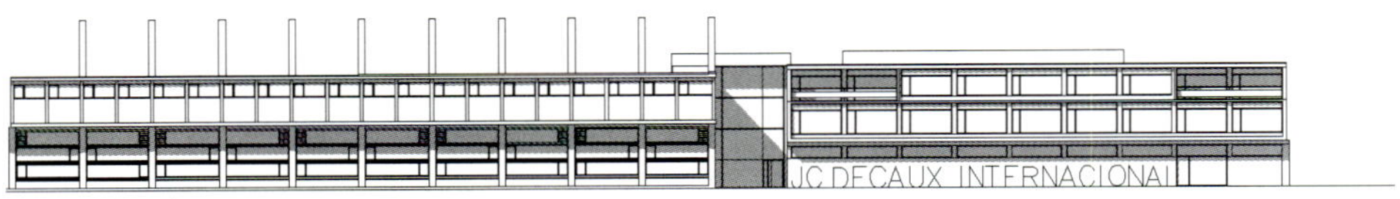

North elevation

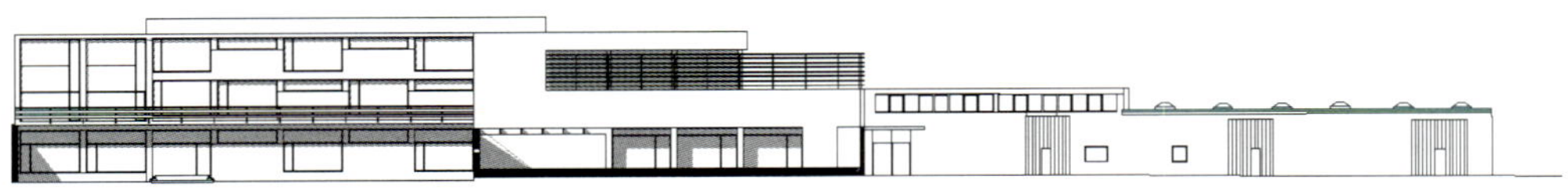

South elevation

The space for receiving clients and visitors is a small conference hall and a double height cafeteria that opens onto a courtyard-garden.
The large rear garden forms a more intimate space than that of the offices.

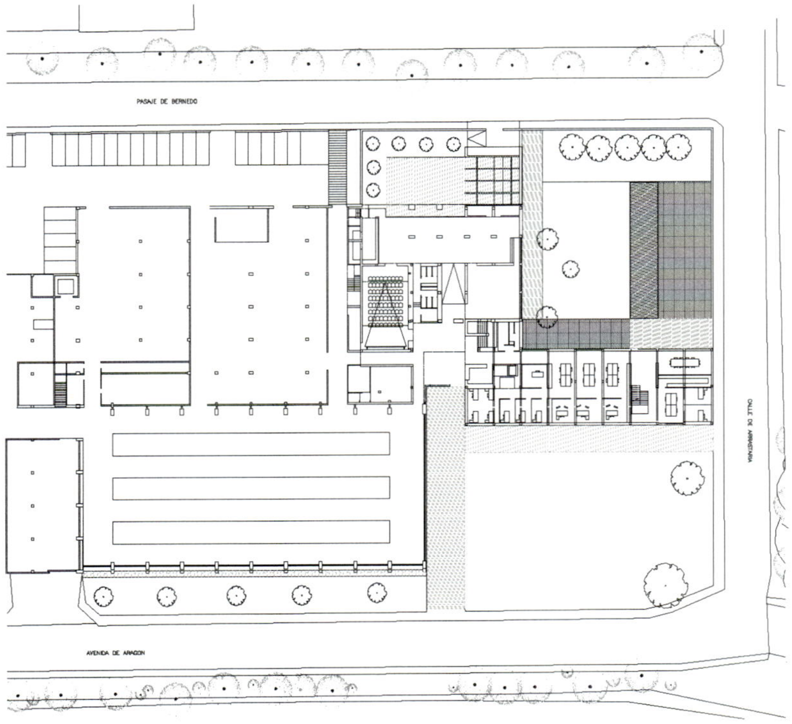

Ground floor plan

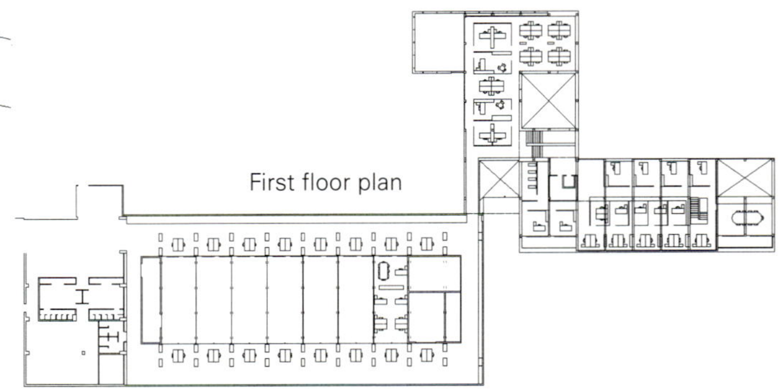

First floor plan

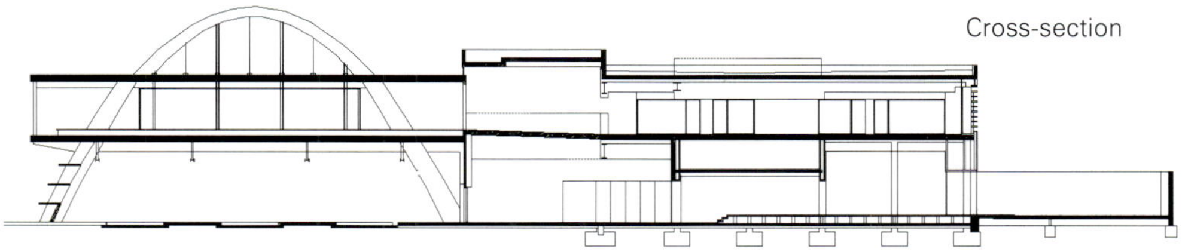

Cross-section

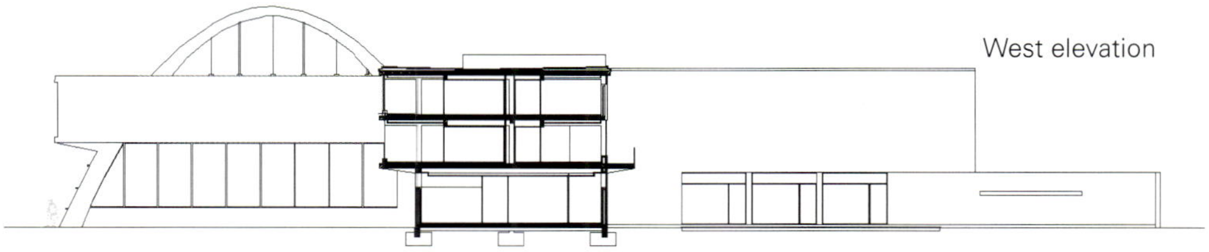

West elevation

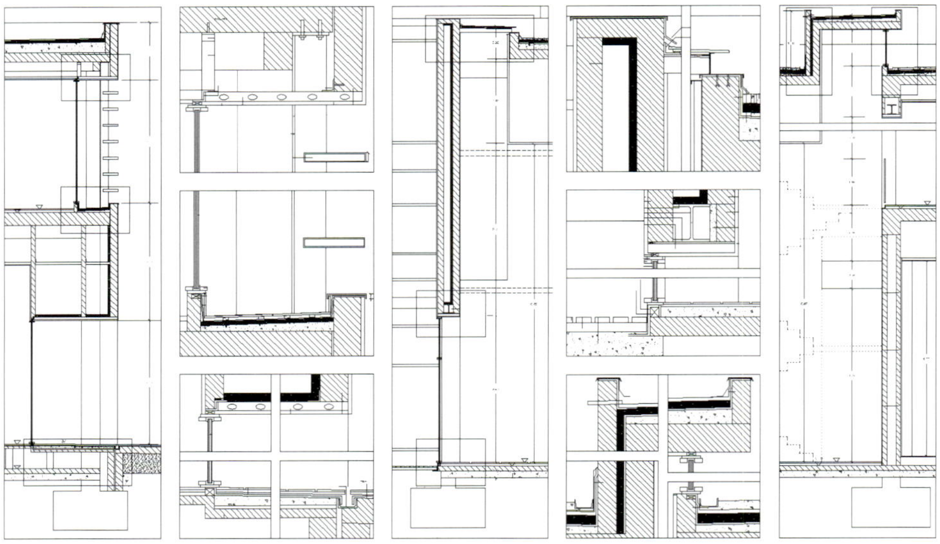

Construction details

The new access breaks into a triple height foyer, the nerve-centre of the communications of the building that provides access to the offices that occupy the original administrative building of the large covered building. The latter has a structure of arches (the former bottling shop of Martini & Rossi) and has been converted into a showroom of urban furniture.

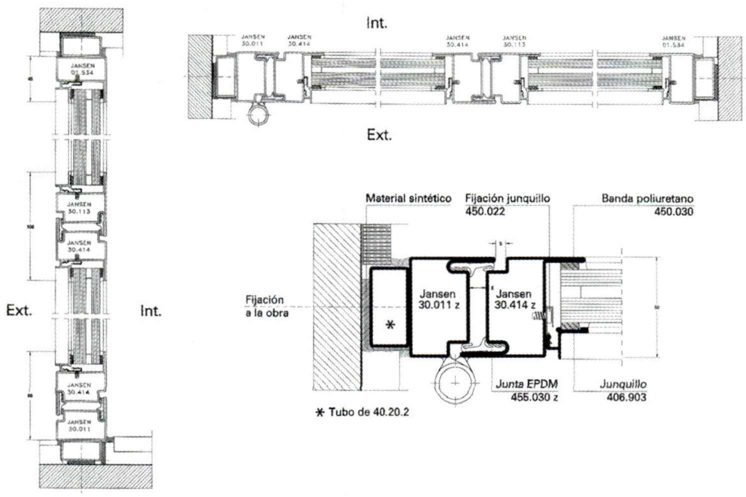

Construction details of the framing

Correa+Estevez Arquitectos

Sede Polígono de Granadilla

Tenerife, Spain

Photographs: Correa+Estevez Arquitectos

The Granadilla Industrial Estate is located on the south-east coast of Tenerife and forms a great industrial area linked to the future port of the island. The brief was to create a multipurpose building to house the offices of the company that manages the estate.

The choice of the plot was decisive in the general approach to the intervention. The building is set on a platform located on an existing concrete dyke in a ravine listed as a zone of interest from which the water has been diverted. The building can be seen from the motorway and the access zones.

To protect the complex from the prevailing winds, the building was surrounded with thick stone walls that enhance the landscape and isolate it from the hostile environment. A microlandscape is thus created inside it with a carpet of vegetation, a sheet of water, a euphorbia forest, and volcanic tufa of different sizes and textures forming a dry garden presided over by a sculpture. The porticoed walkways define an itinerary that is intended to be surprising and pleasant for the spectator. As the different modules of the complex are supported on a flat horizontal base, the land flows freely beneath them. The volumes have been arranged through an interplay on either side of the longitudinal axis created by intercalating two different treatments: cubicles enclosed by stone walls, and light metal structures adjoining the corridor.

The brief called for a tower-belvedere from which the plots could be viewed and which would act as a vertical counterpoint to the extremely linear buildings. It was therefore located at the beginning/end of the walkway, with structural continuity and a similar formal treatment.

The centre also has a guard module, a glazed box that houses the open-plan offices –protected from the sunlight by a pergola that surrounds it–and a functions room at the end of the trajectory that is enveloped by a stone wall.

The arrangement of the different elements allows another office module to be added parallel to the existing one for a future extension.

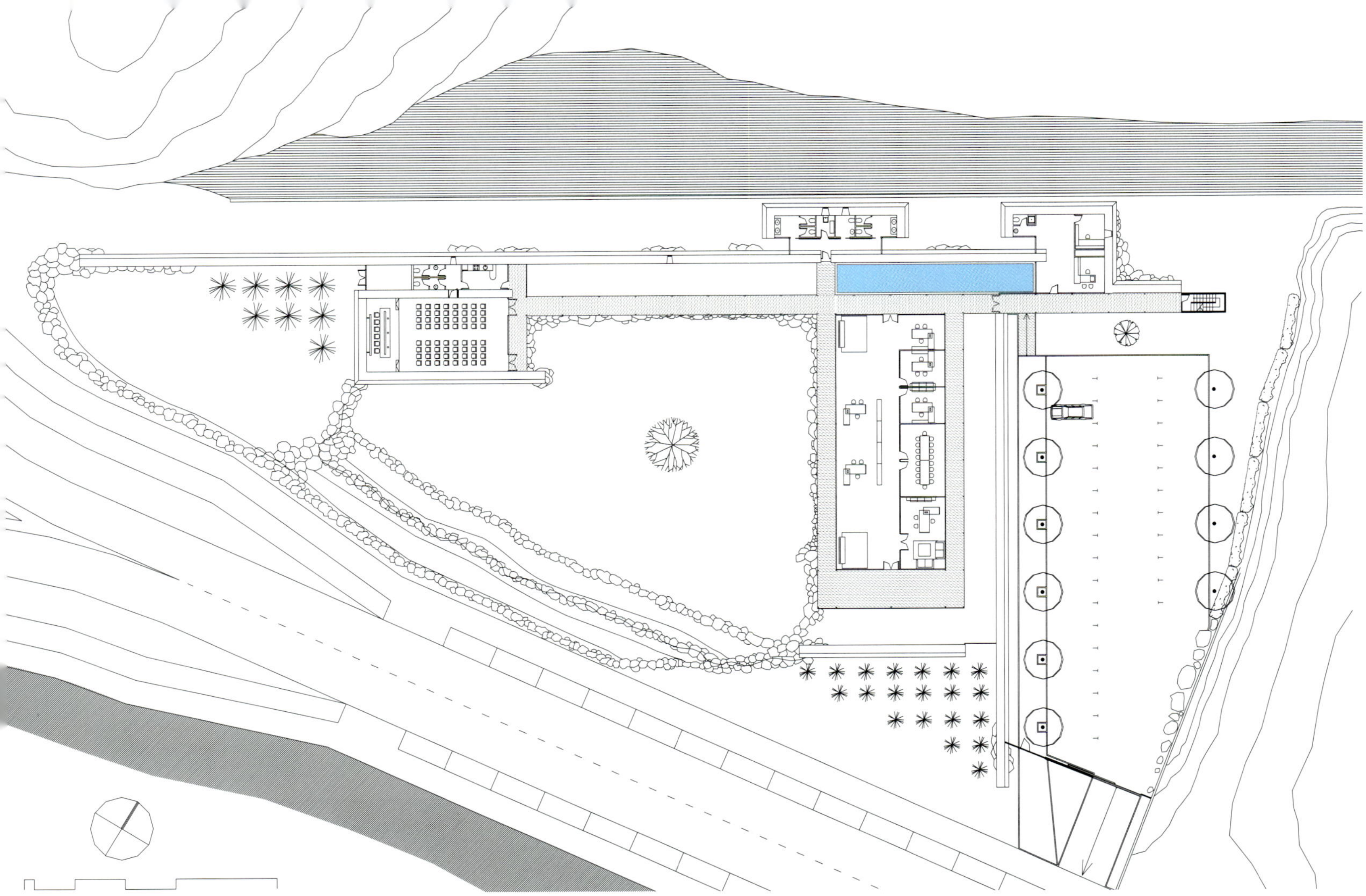

Floor plan

Cross-section

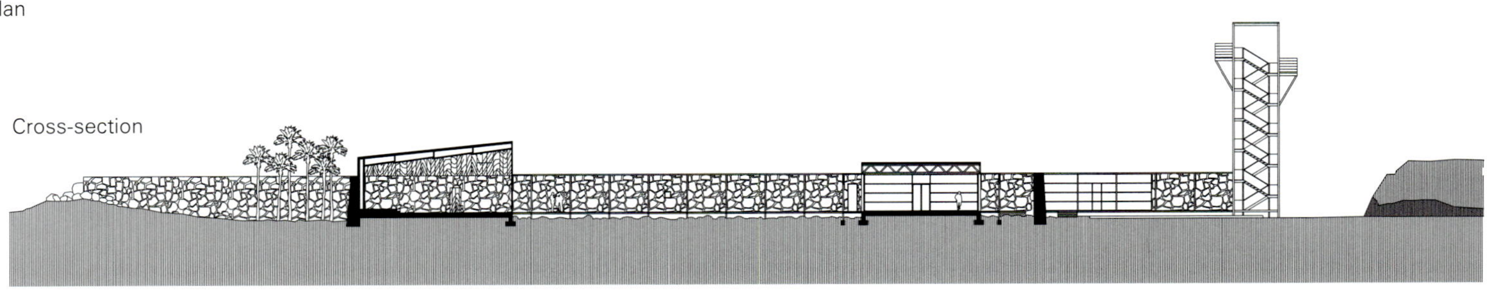

The scheme was designed bearing in mind the location of the building, the desert landscape, the continual wind that blows obliquely along the coast, and the strong sunlight.

North elevation

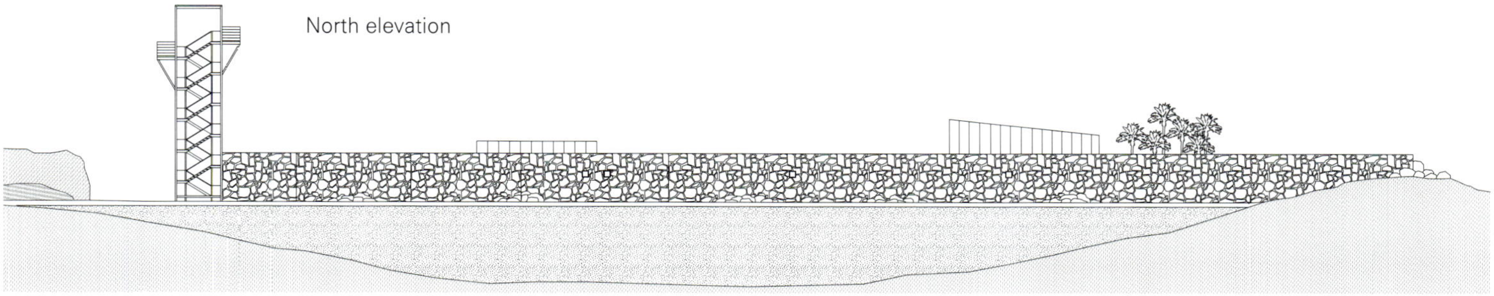

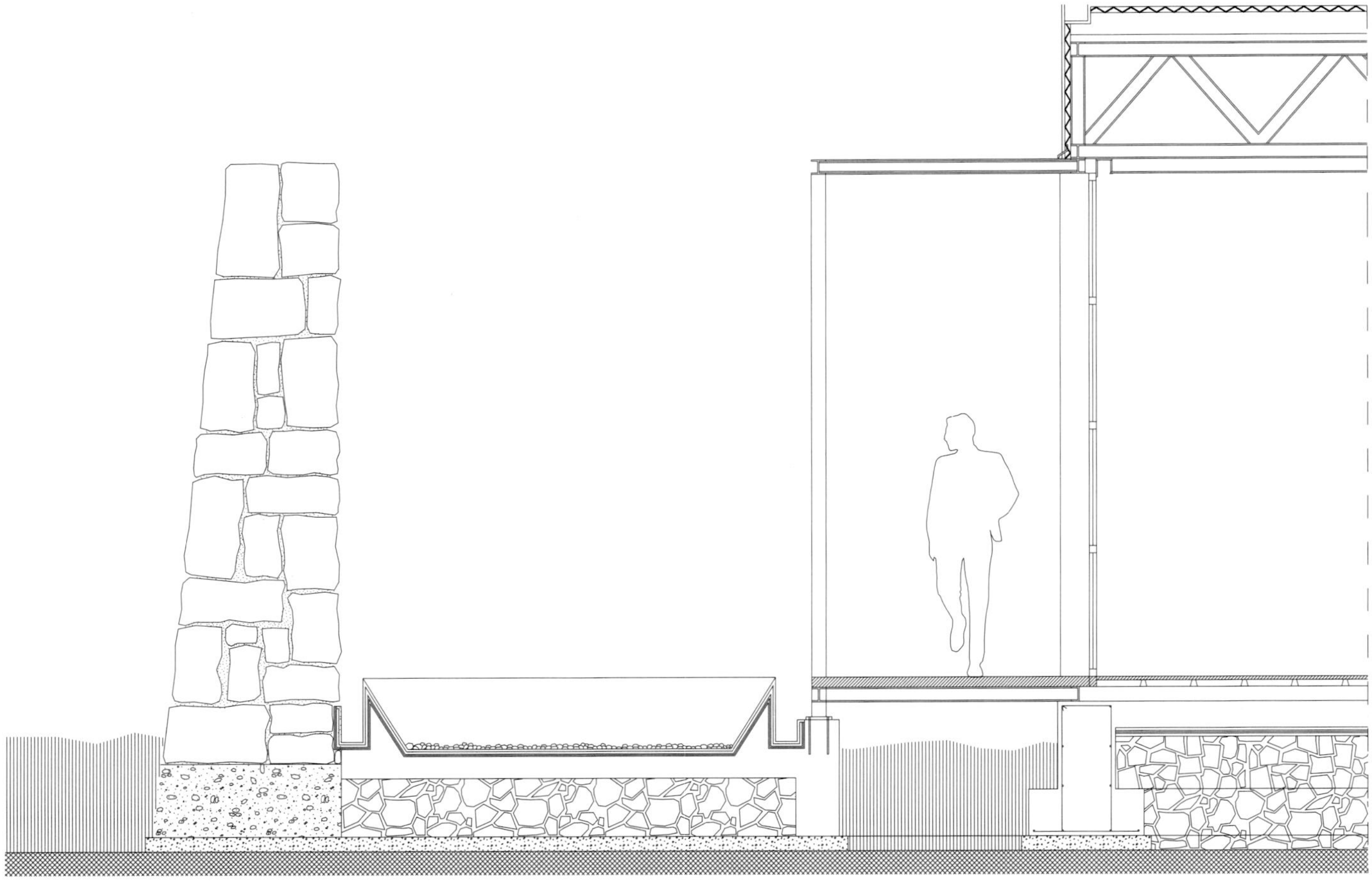

Construction sections

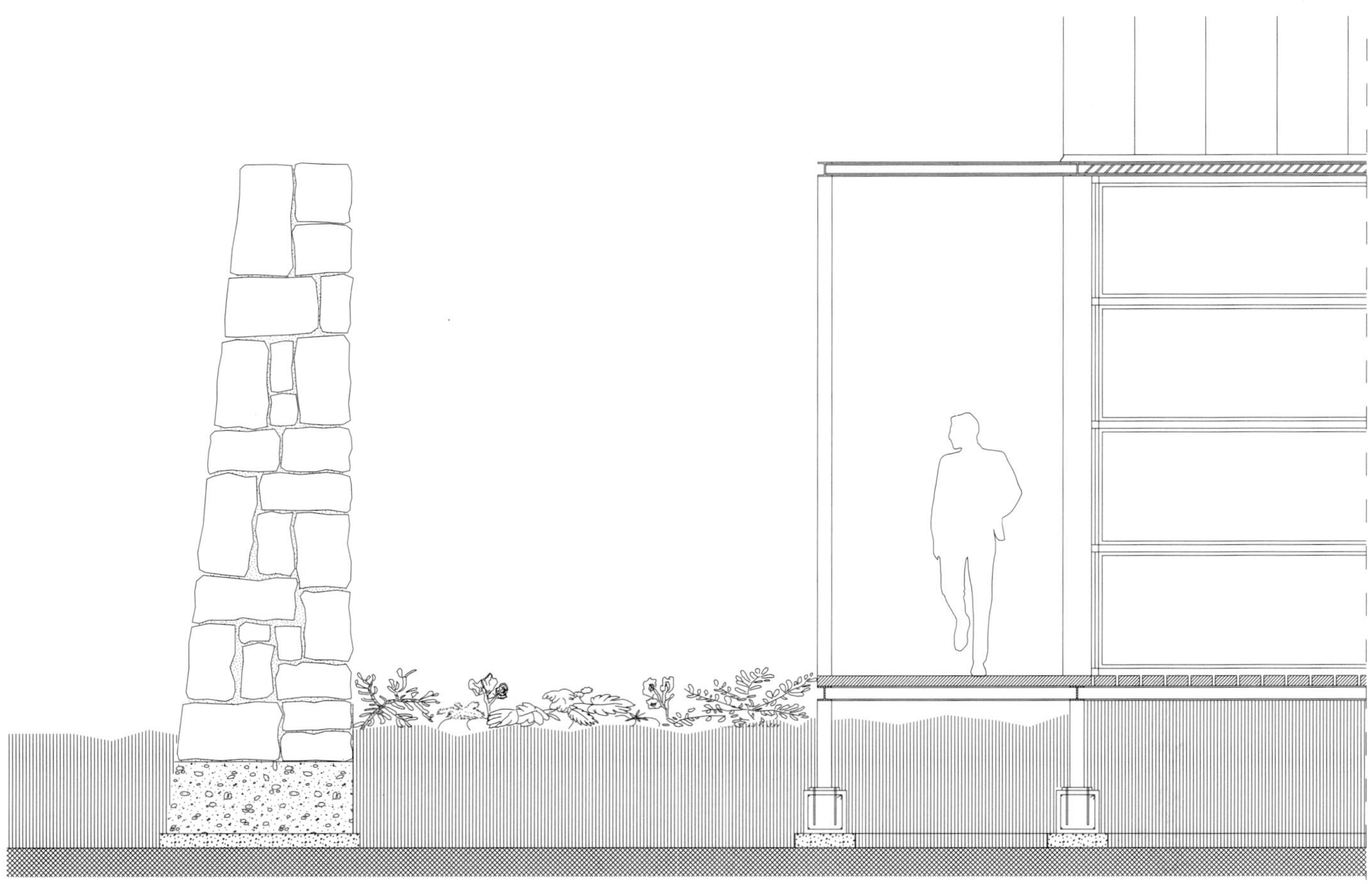

The materials are defined on the basis of the landscape. The structural walls of rubble work are left bare on the interior, thus providing continuity with the exterior, whereas the internal separations consist of cedar wood partitions.

Florian Nagler Architekten

Factory Hall

Bobingen, Germany

Photographs: Stefan Müller-Naumann

Kaufmann Holz AG needed a simple warehouse with cranes for the storage, finishing and commissioning of glued laminate elements. A building with a suitable, simple structure and details was to be built under enormous time pressure.

A transparent facade made from impact-resistant, full-height polycarbonate sheeting means that the activities being carried out within the warehouse —workers, production stages, materials, construction— can be viewed from the outside, while also allowing the surroundings to visually filter directly into the hall. This treatment of the facade also means that it changes appearance, depending on the viewpoint, weather and lighting conditions. Since a series of gates were to be placed side by side along the facade, they have been arranged so that they open with a vertical sliding movement, thereby ensuring that these massive doors would be as unobtrusive as possible.

Glued laminate battened columns, whose deep cross section creates a fixed joint, create a yoke-like two-bay hall. The inner columns take up the load of the crane gantries and the outer columns support the loads from the roof structure.

The catwalk for observation and driving of the five cranes in both bays runs through the central columns. As the primary structure over the individual bays, slender glued element beams at two meter centers were selected independent of the timber structure. Temporary steel platforms have been installed in the area of the full-height saw and planning machines.

The resultant building is not only extremely functional, but is of one piece with regard to materials, construction and detailing. The client is not only satisfied with the functioning and appearance of the building, but also with the "Corporate Identity" image that the building conveys of this timber construction firm.

The transparent facade is made from impact-resistant, full-height polycarbonate sheeting. The gates have been arranged so that they open with a vertical sliding movement in order to make them as unobtrusive as possible.

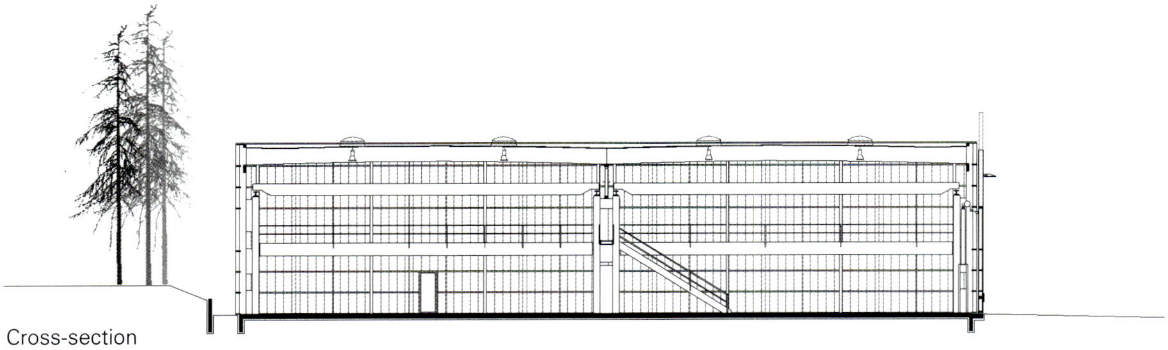
Cross-section

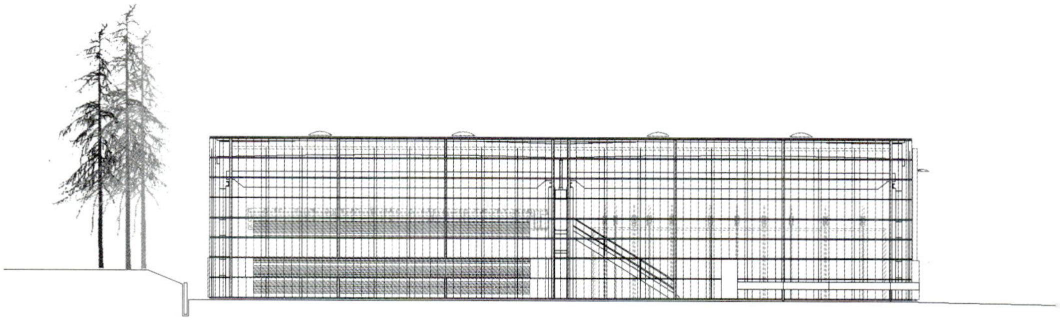

West elevation

The catwalk for observation and driving of the five cranes runs through the central columns. Slender glued element beams at two meter centers, form the primary structure over the bays.

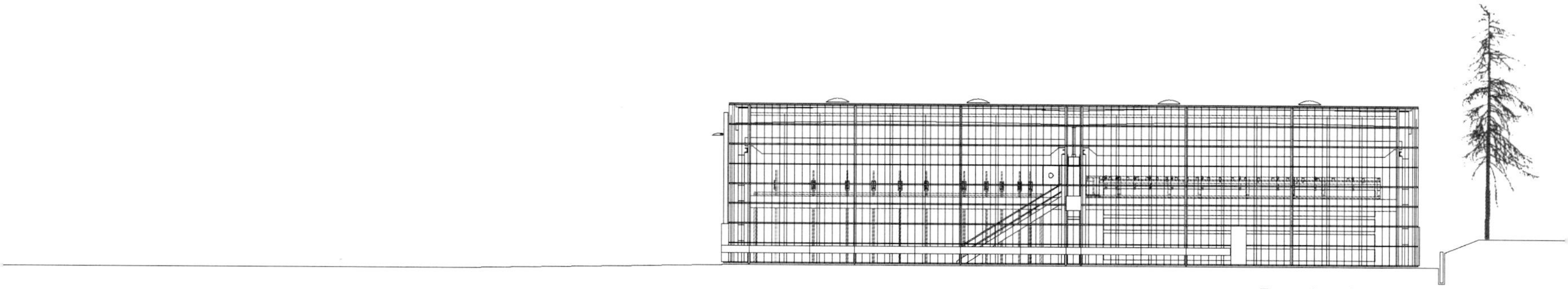

East elevation

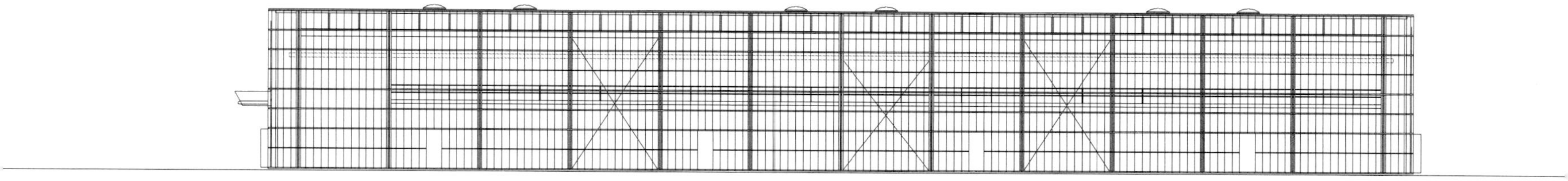

North elevation

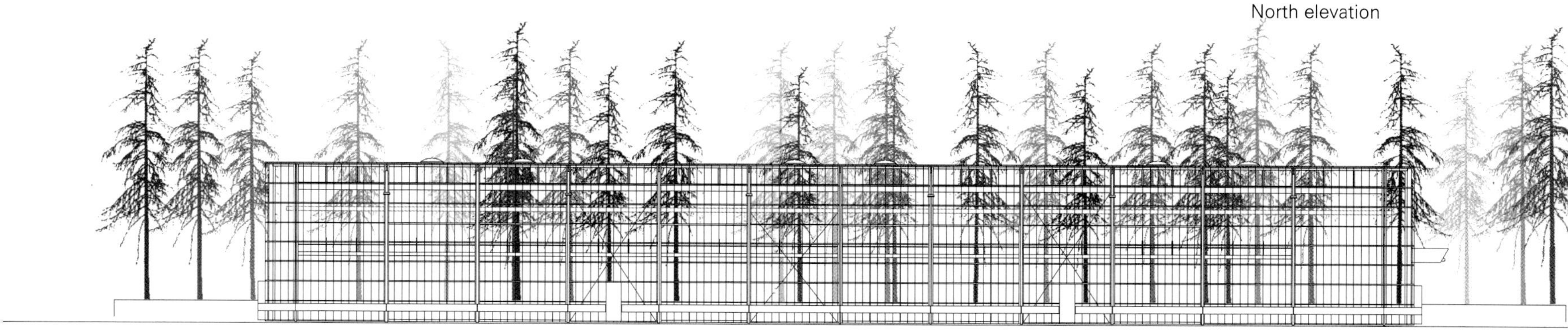

South elevation

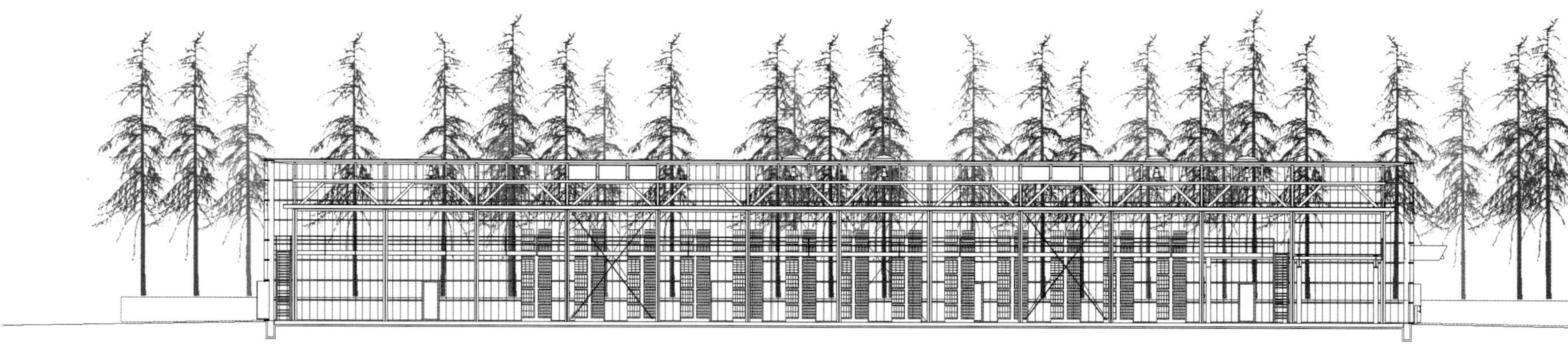

Cross-section

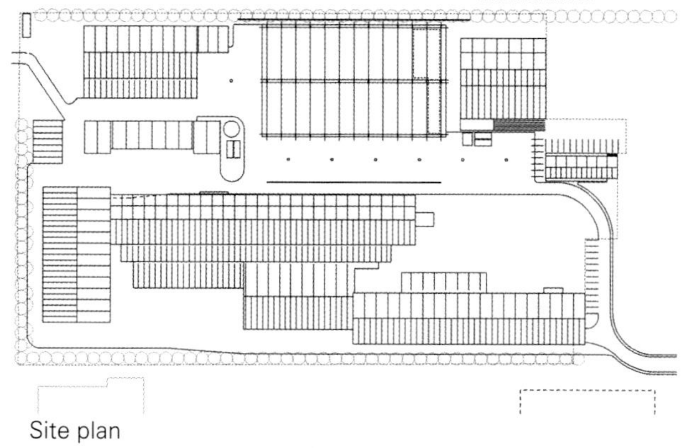

Site plan

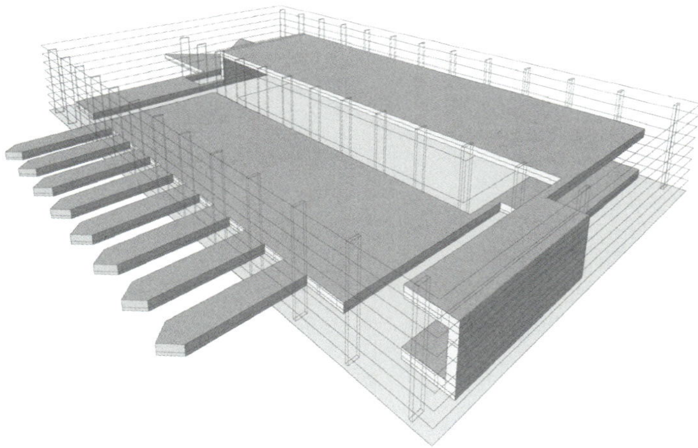

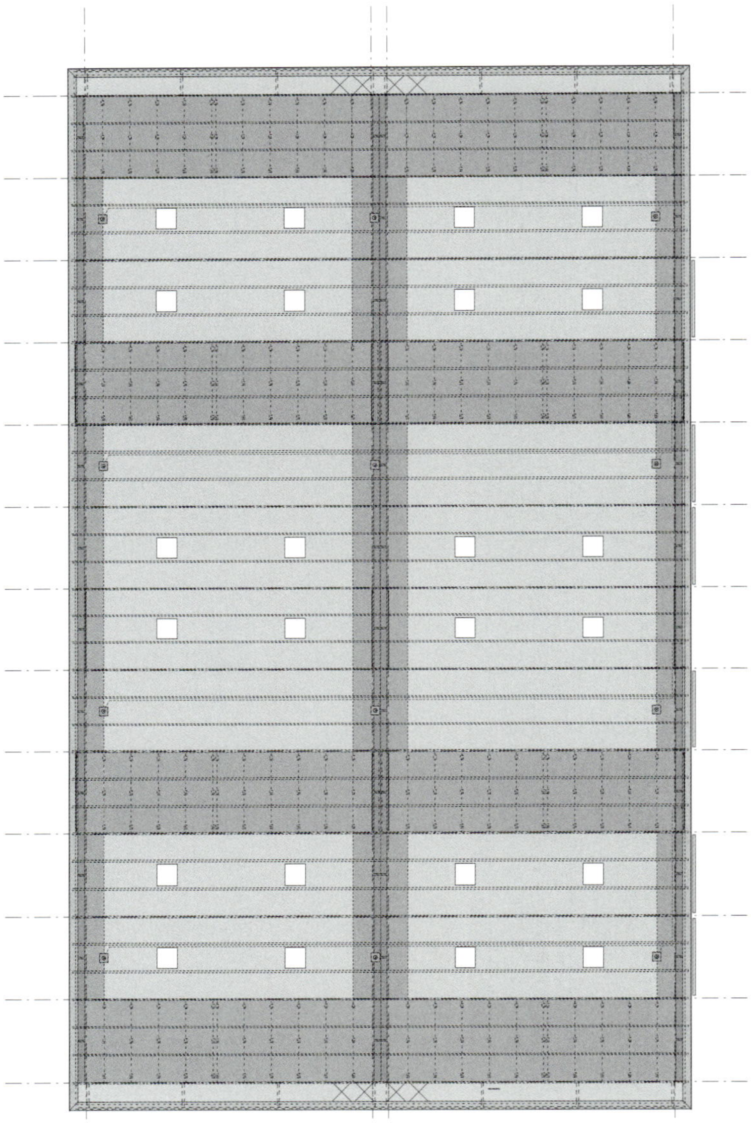

Glued laminate battened columns, whose deep cross section creates a fixed joint, create a yoke-like, two-bay hall. The inner columns take up the load of the crane gantries, and the outer columns support the loads from the roof.

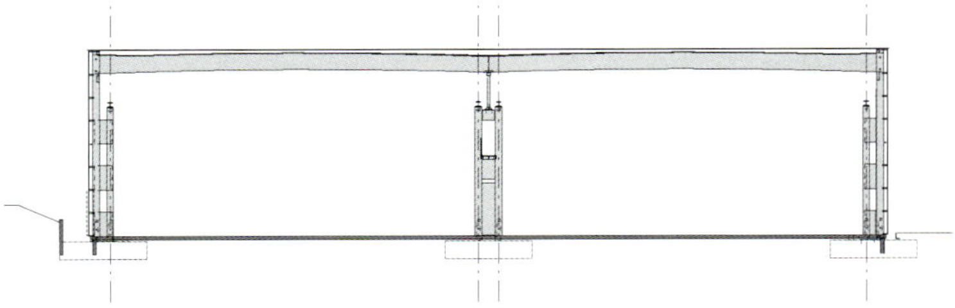

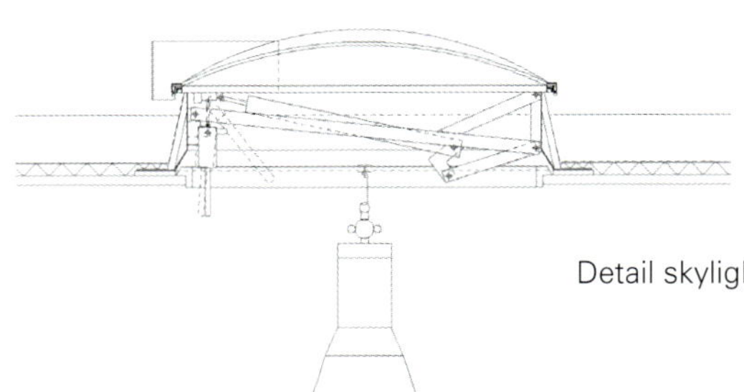

Detail skylight

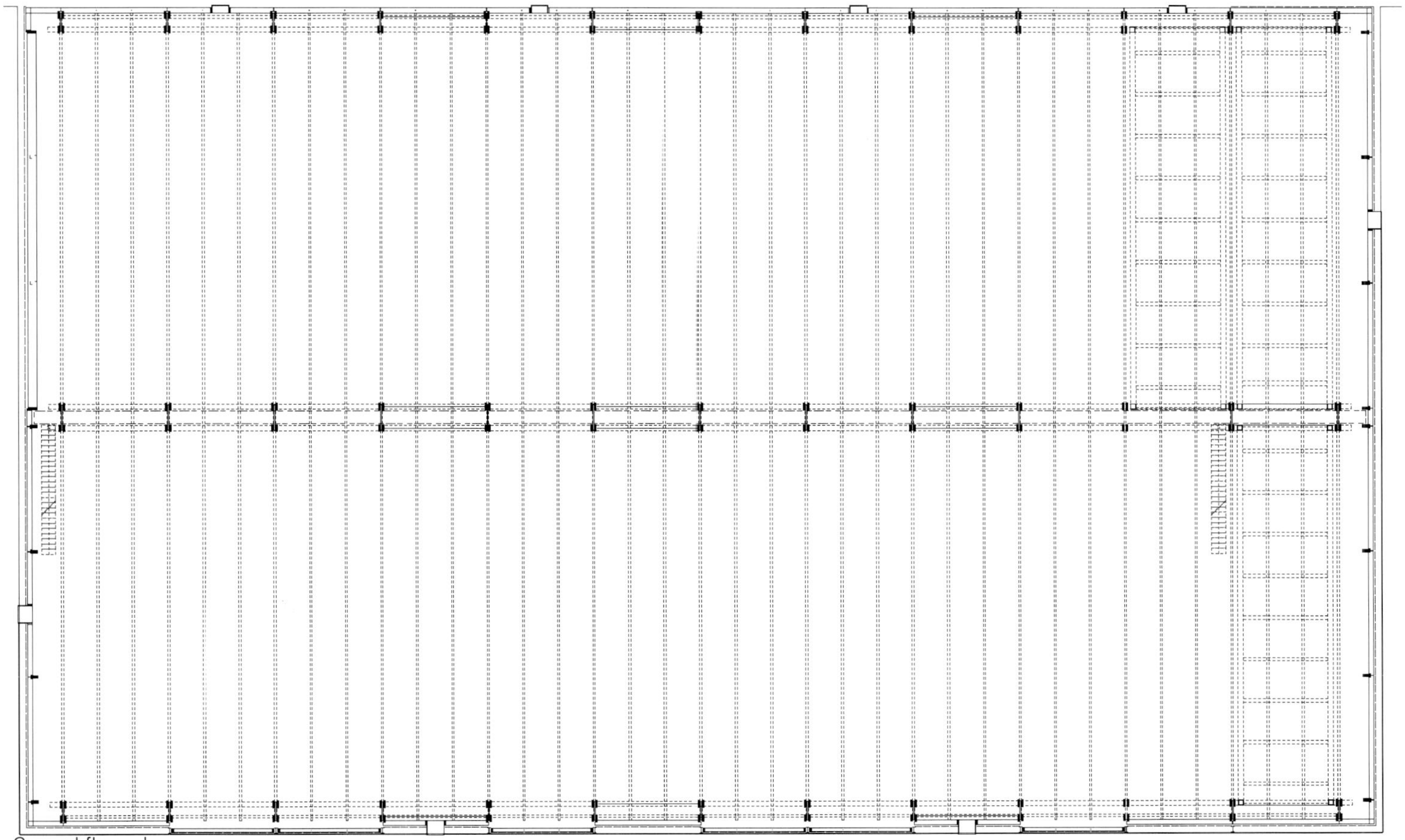

Ground floor plan

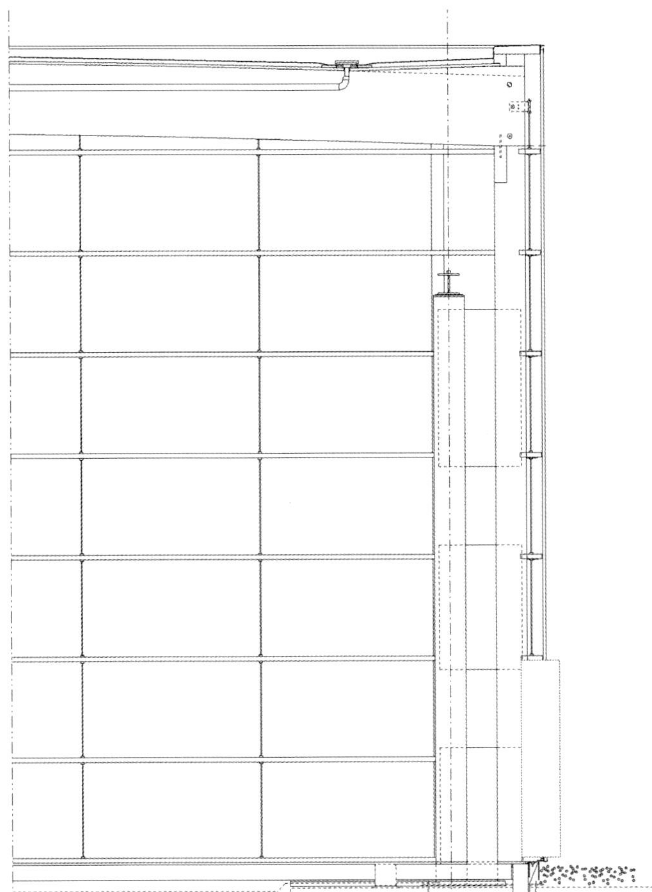

Facade section

Detail gate

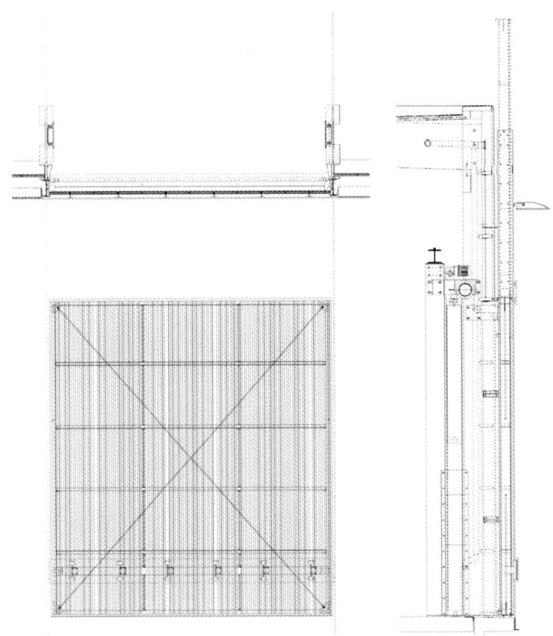

Henn Architekten

Auto Türme

Wolfsburg, Germany

Photographs: Werner Huthmacher

With the delimiting factor of the highway lying to the north, a long, rectangular pond has been constructed to the northwest, out of which rise two steel and glass cylinders. Four more of these cylinders are also slated for future construction. These towers, which function as elevated storage space for vehicles before they are sent to the dealership, symbolize the end result of the production process.

Each tower has a storage capacity of 400 vehicles, which are moved to a free space on one of the twenty stories via a computer-controlled elevator. A new vehicle makes its arrival –by way of an underground route– every forty seconds, just as another leaves the tower, en route to a car dealership. This continuous movement within the cylinders symbolizes the constant rhythm of a highway, as if comprising some sort of transparent engine.

In serving their function as raised warehouses, the towers have been equipped with a galvanized steel structure, with supports arranged along the radii. Cantilevered arms, emanating from the supports, bear the weight of the facades main sections. At the same time, this structure has enabled a service walkway along the perimeter of the tower. A central elevator moves the vehicles to the storage slots, which consist of prefabricated metal panels arranged on the radii of the structure steel structure.

The physical distribution of the cars, from the production site to the sales-floor, with provisional storage along the way, is all done via a conveyer belt in an underground tunnel. With a system of guides and platforms, the vehicles are introduced into the glass towers and deposited, via one of the four elevators, onto the corresponding platform. Each elevator covers the slots for half a tower.

The new vehicles are tested once again in the tunnel before a conveyer belt moves them to the delivery point. The height of the conveyer belt is adjusted to the same height of the upper edge of the platforms, and the vehicles now roll for the first time on their own wheels.

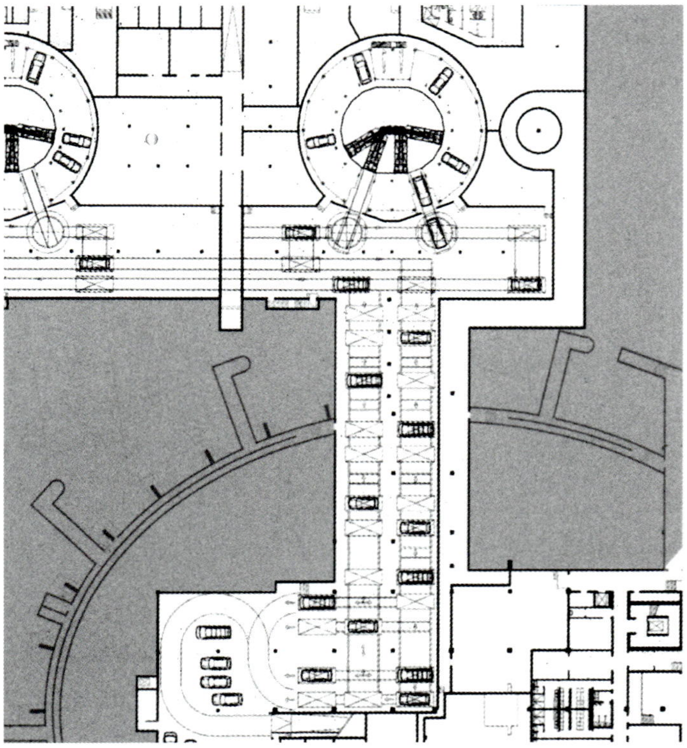

Plan of the cars distribution

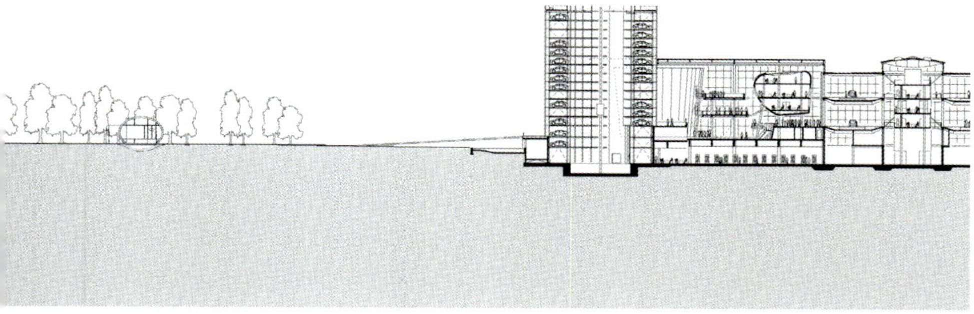

The towers have been equipped with a galvanized steel structure, with supports arranged along the radii. Cantilevered arms, emanating from the supports, bear the weight of the facades main profiles.

Cross-section

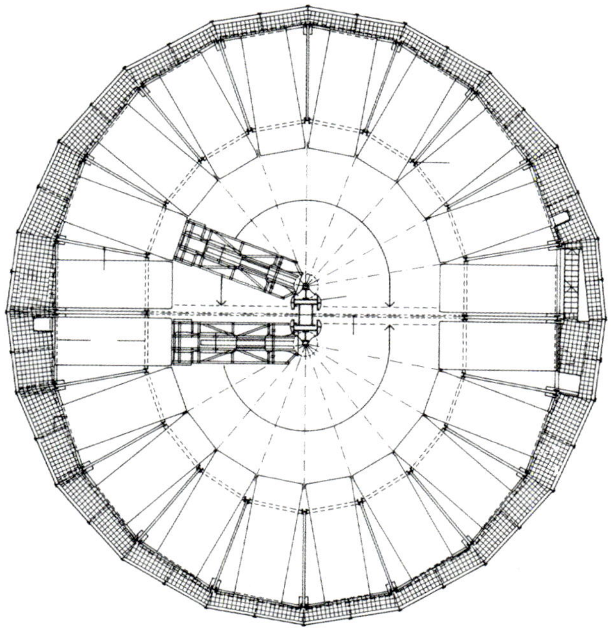

Type floor plan

Henn Architekten

Gläserne Manufaktur

Dresden, Germany

Photographs: Werner Huthmacher

11 December, 2001, marked the official opening of the Gläserne Manufaktur Dresden located on the edge of Dresden's historical center. In the near future the complex's spacious interior will house the vehicle assembly and customer center for Volkswagen's recently developed Phaeton Limousine.
In contrast to ordinary distribution centers the Gläserne Manufaktur Dresden celebrates not only the handing over of the vehicle to the customer but, above all, its production processes. Thanks to the transparency of the building, the step-by-step assembly of the vehicle is visible even from the outside, as it is manufactured primarily by hand in the bright studio-like interior. The punctual delivering of the components is guaranteed by the specially designed Cargo Tram, which uses the existing public transport track layout. Before the bodywork of the car enters the assembly line, it is presented showroom-like in the glazed body-warehouse, which is visible from the street. In the same way the finished models are stored in the completely transparent car tower before being handed over to the customer.

The building volume between the assembly and the public areas houses exhibitions, simulators, restaurants and bars, thus completing the palette of services and events on offer. Not only can the customer participate in the production of this vehicle, the people of Dresden can also participate in the process as an urban event, in the sense of a market place in the city's public open space.
The unusual combination of topics on show, from technology to culture will enable the Gläserne Manufaktur Dresden to serve as a catalyst for a new appreciation of the city of Dresden. Simultaneously, different aspects of Dresden's identity are highlighted, going beyond the traditional image of a city of the baroque, its role as an integrated industrial location, its extraordinarily rich cultural and artistic heritage as well as its urban planning tradition as a garden city. Thus the complex creates a space for events utilizing Dresden's cultural reservoir of historical sights and complementing the city space with a modern attraction. The Gläserne Manufaktur will enhance the city's identity.

The factory is reached by tram, or car, across bridges leading over lakes and gardens. Its essentially rectangular form is interrupted by the circular glass tower.

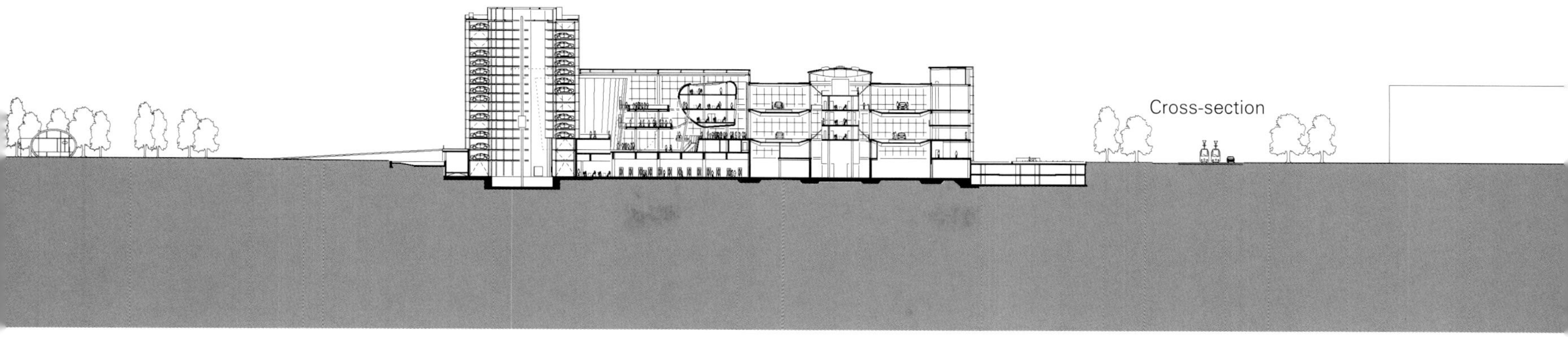

Cross-section

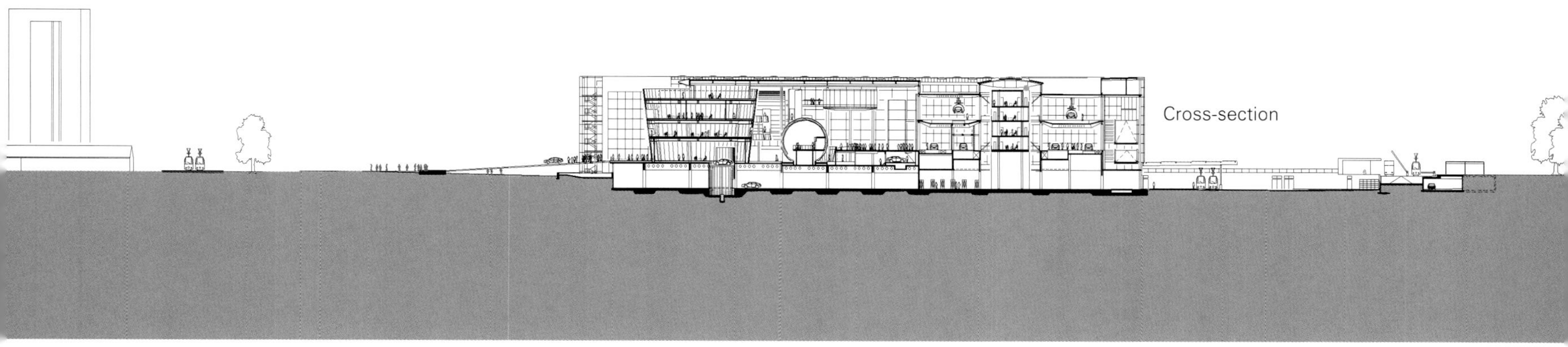

Cross-section

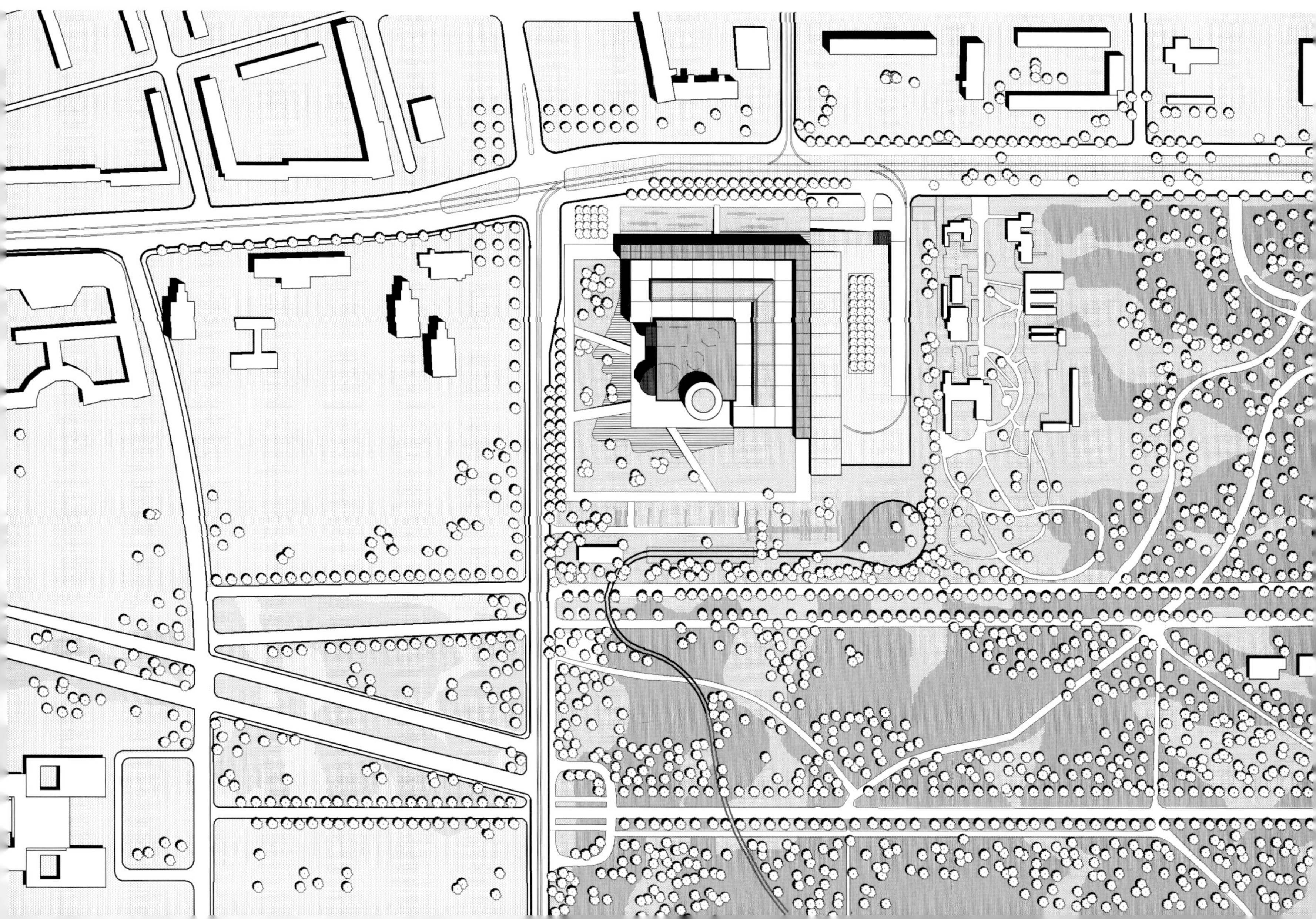

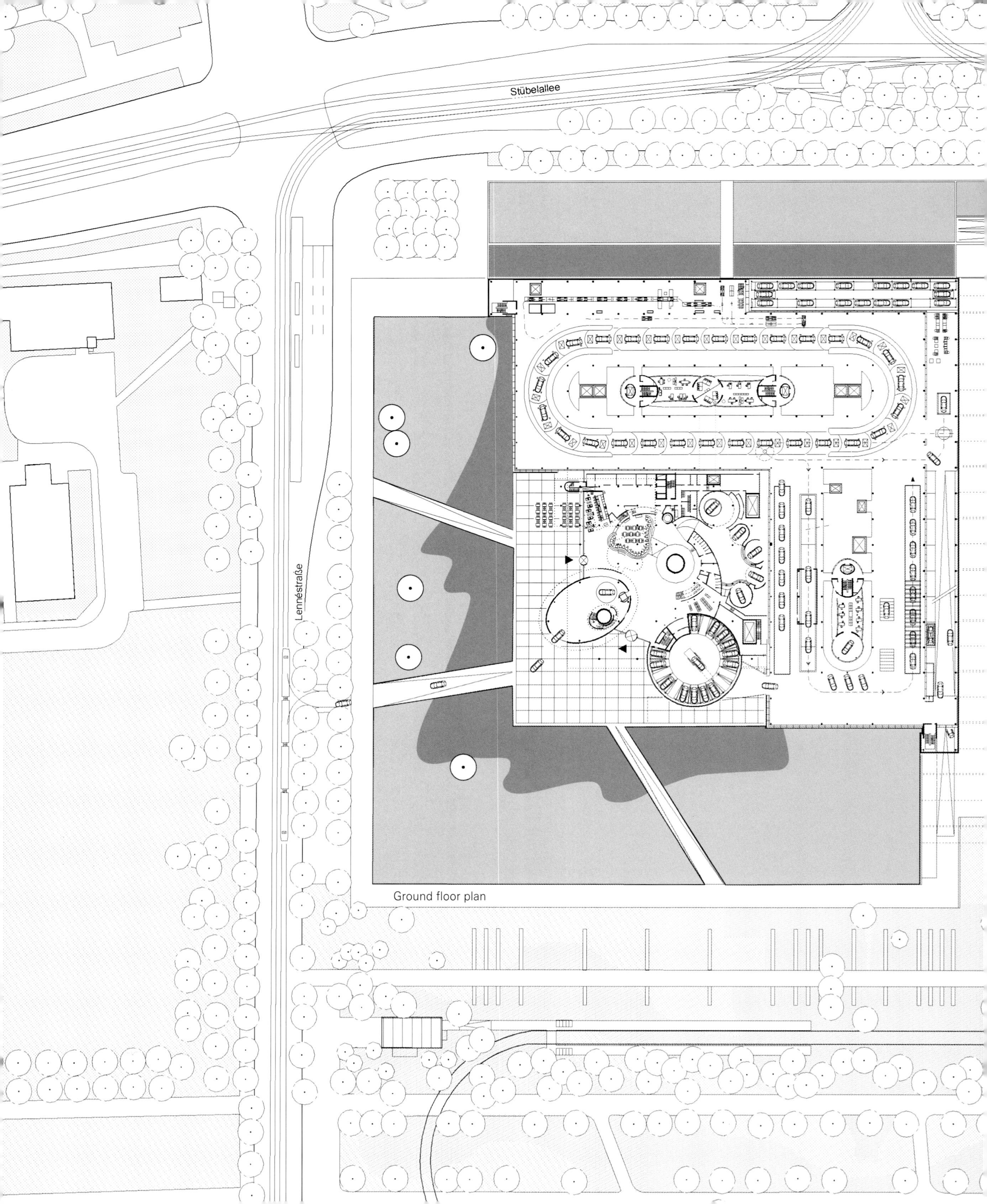

Ground floor plan

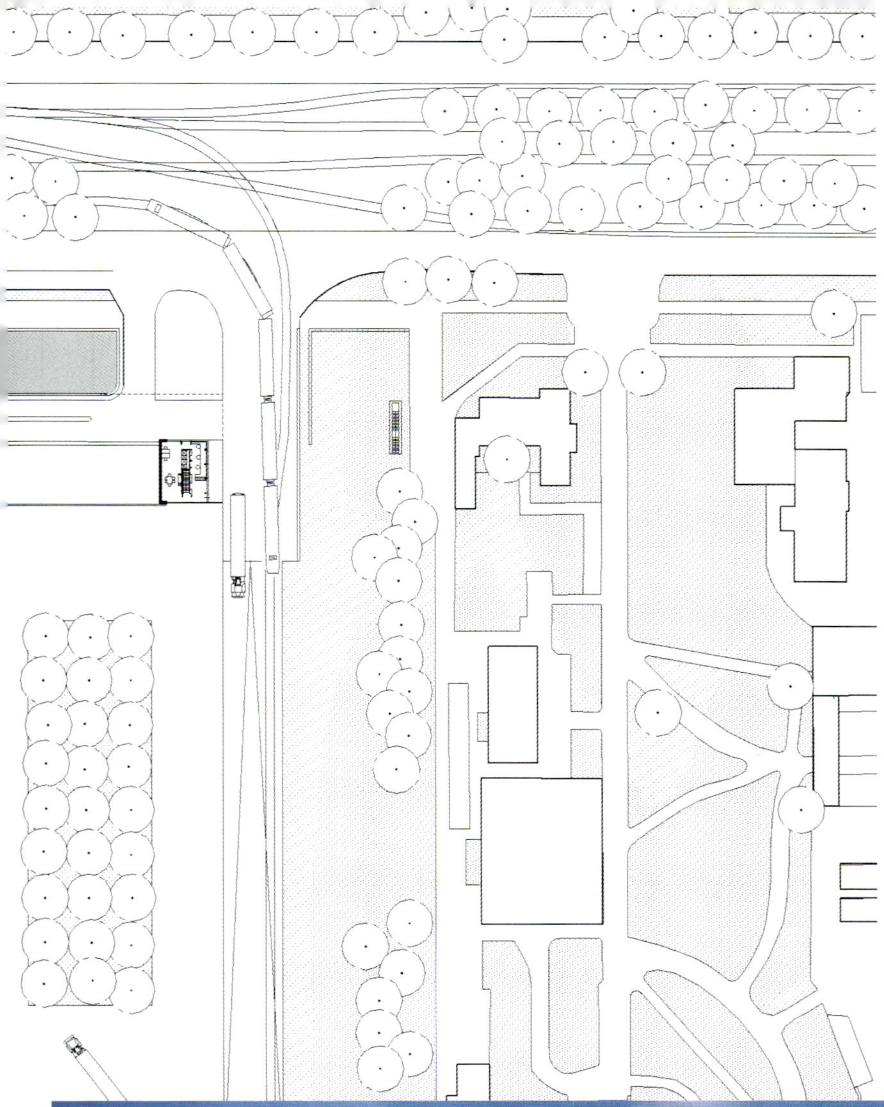

Level 1

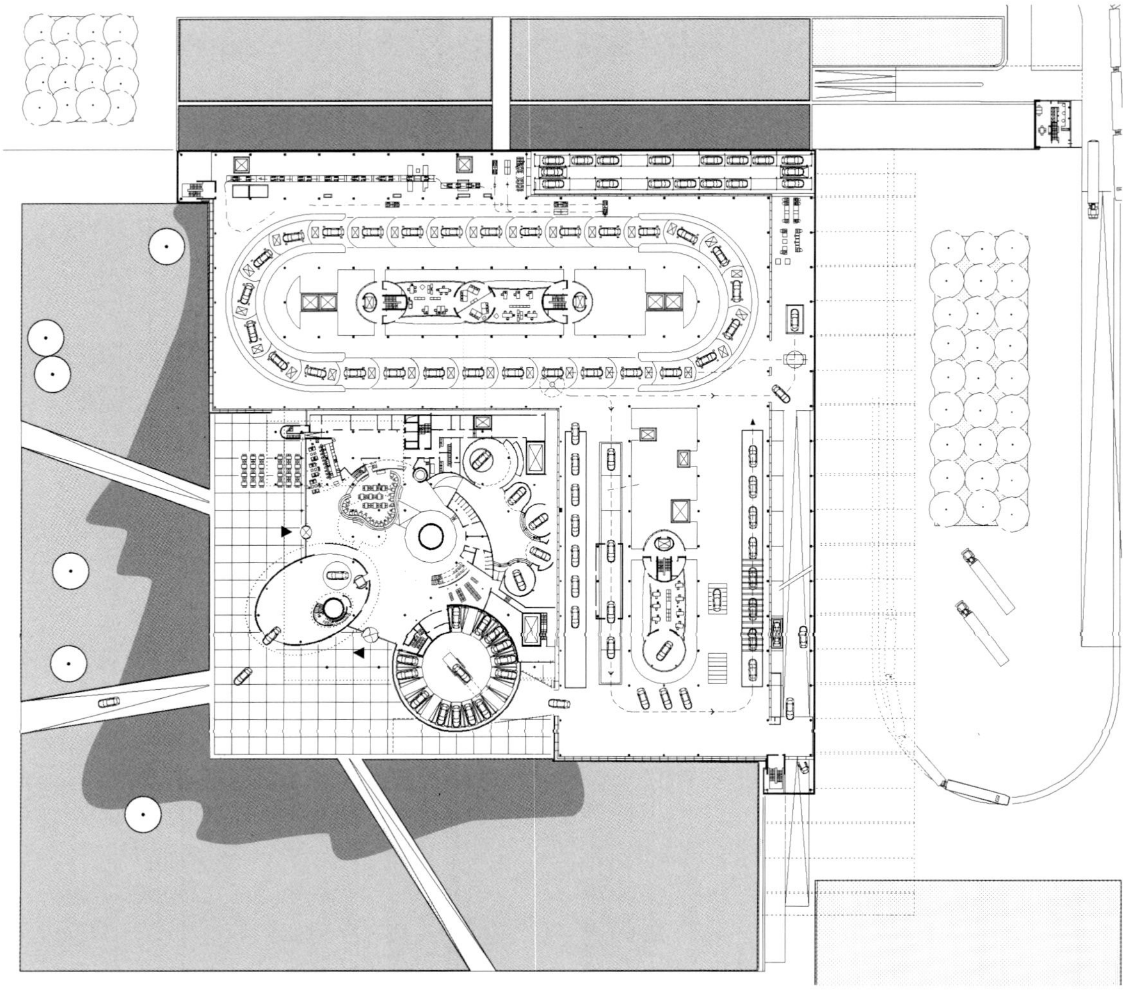

Level 2

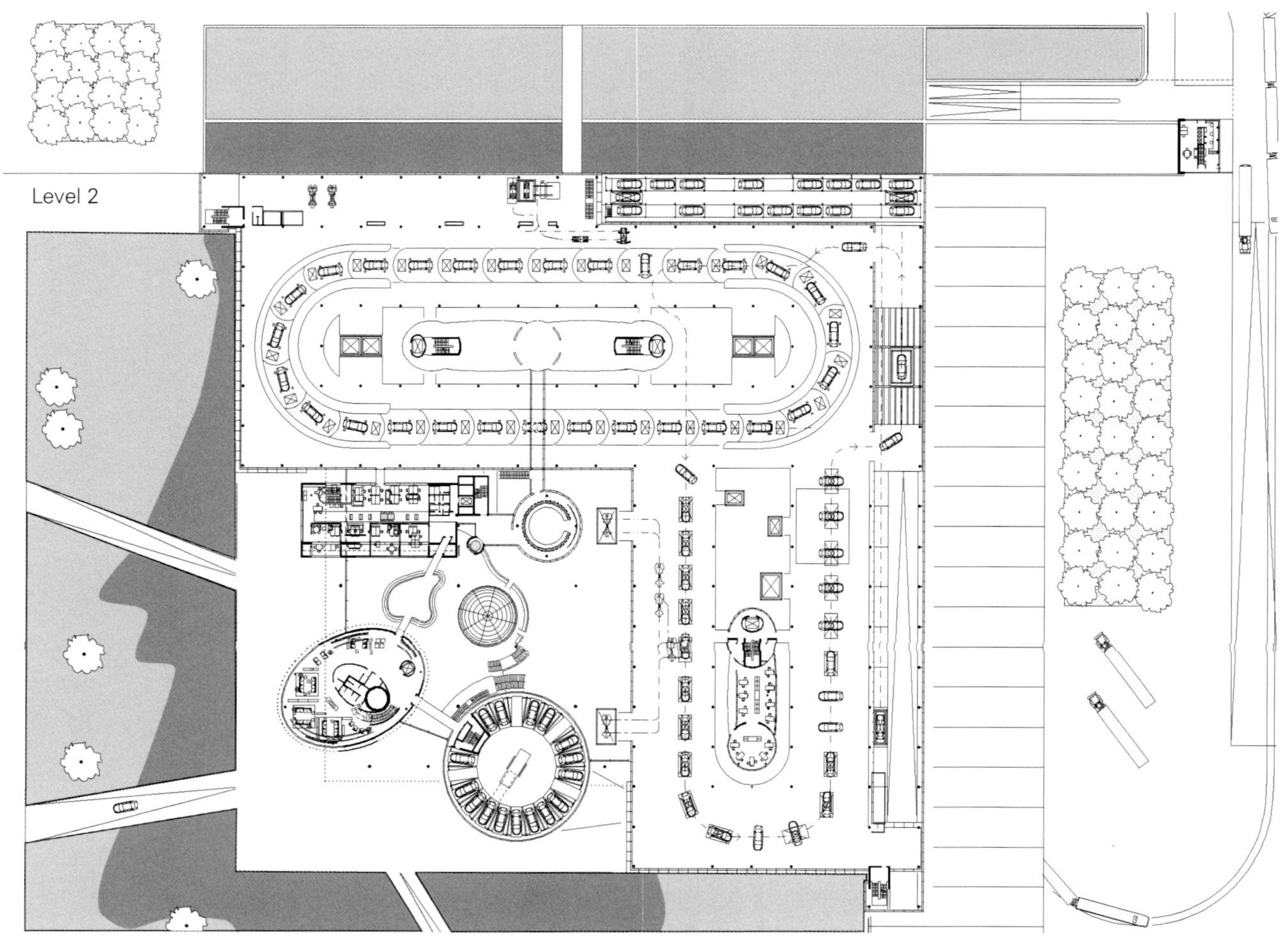

Before the body-work of the car enters the assembly line, it is presented showroom-like in the glazed body-warehouse, which is visible from the street. In the same way the finished models are stored in the completely transparent car tower before being handed over to the customer.

Jacques Ferrier Architecte

Sapeg water treatment plan

Joinville le Pont, France

Photographs: Gaston Bergeret, Jean Marie-Monthiers & Hervé Abbadie

The Marne River water-treatment plant was built by the engineer Belgrand for Napoleon III, and has constantly undergone transformations over time to accommodate changes in the process of filtering drinking water. The end result is an ultramodern plant, marked by the presence of water and hundred years-old trees, has also become precious natural breathing space in the urban context now surrounding it. The project preserves and enhances this characteristic: designed as a horizontal layer of alternating ponds and planted terraces, the new plant looks more like a landscape or a work of art than a building.

The project is organized on an even grid which extends out form that of the existing filter beds. Ten 40-metres squares are lined up on either side of a technical gallery and the functional units are freely distributed on this grid; filters and technical equipment have been able to evolve during the project without changing the homogeneity and coherence of the whole. The architecture, like Louise Nevelson sculpture or a Le Nôtre garden, structure the diversity of the areas in this open-ended industrial facility by giving it an enduring formal order that guarantees a unified, consistent image.

All of the blocks are of the same size and height, and covered with large panels of a blue concrete, embedded with turf and the filters have been left uncovered where the sound of constantly flowing water can be heard, whereas in the central nave below, technology is put on display by exposing the powerful water pipes.

All of the blocks are of the same size and height, and covered with large panels of blue concrete, embedded with vitreous glass fragments.

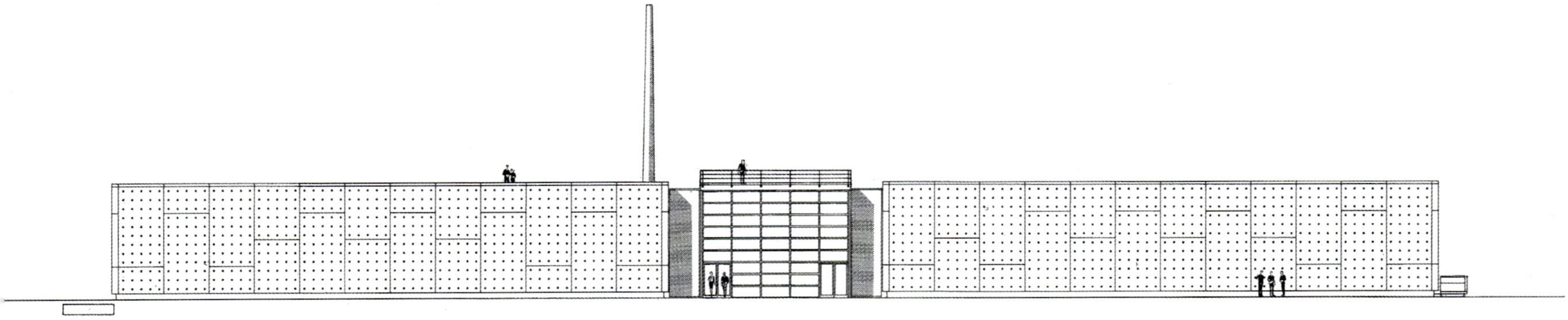

East elevation

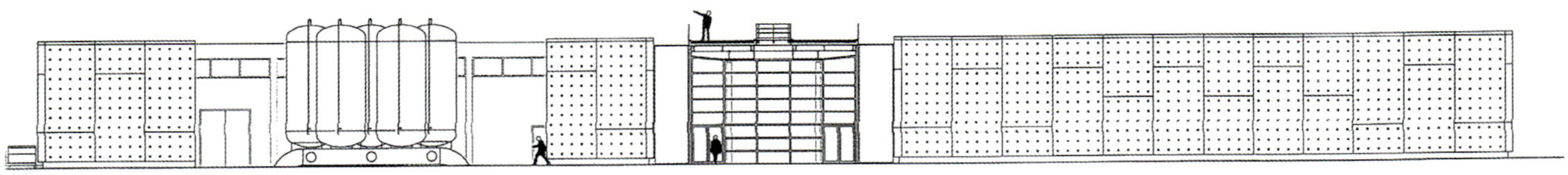

East elevation from the gallery

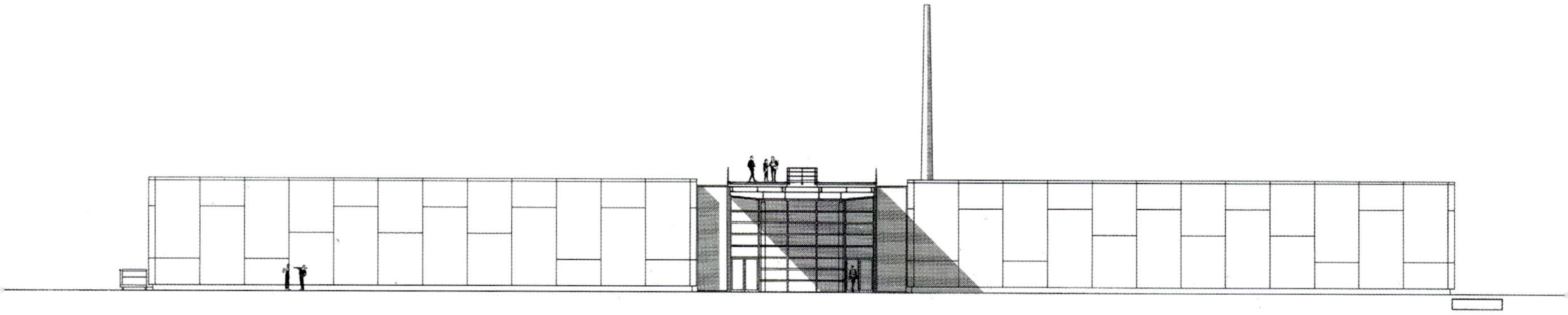

West elevation

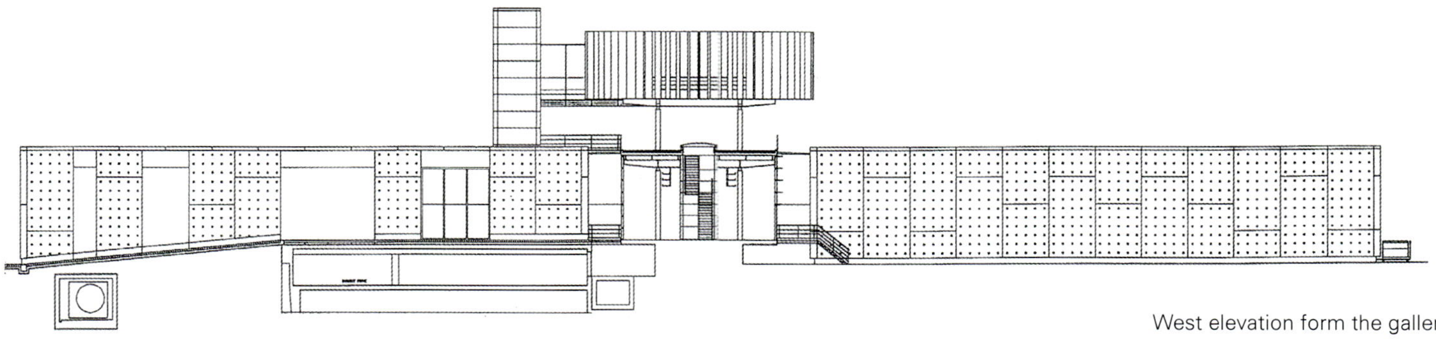

West elevation form the gallery

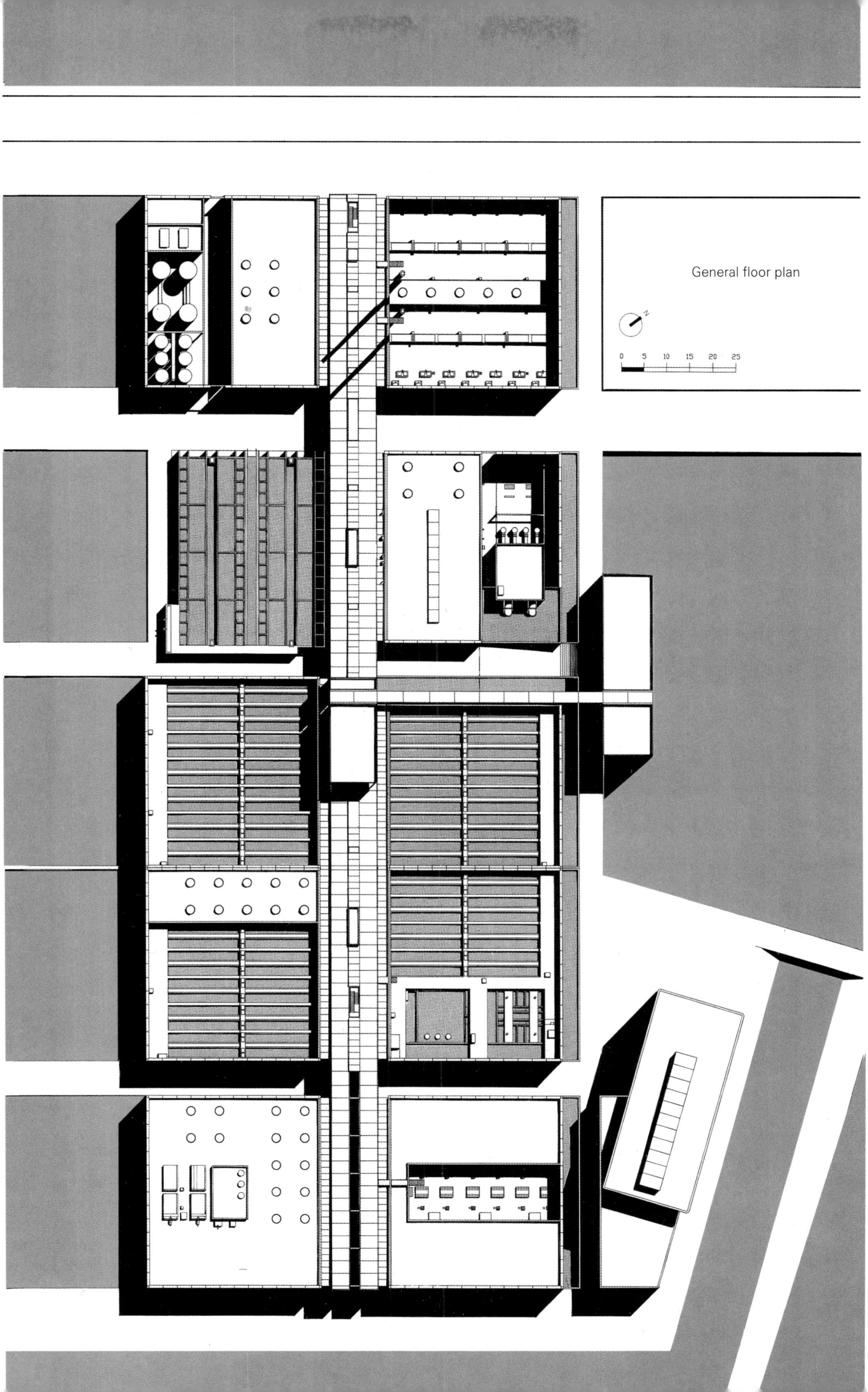

General floor plan

The project is organized on an even grid which extends out from the existing filter beds. Ten 40-meter squares are lined up on either side of a technical gallery and the functional units are freely distribued on this grid.

Roof plan

Gallery cross section

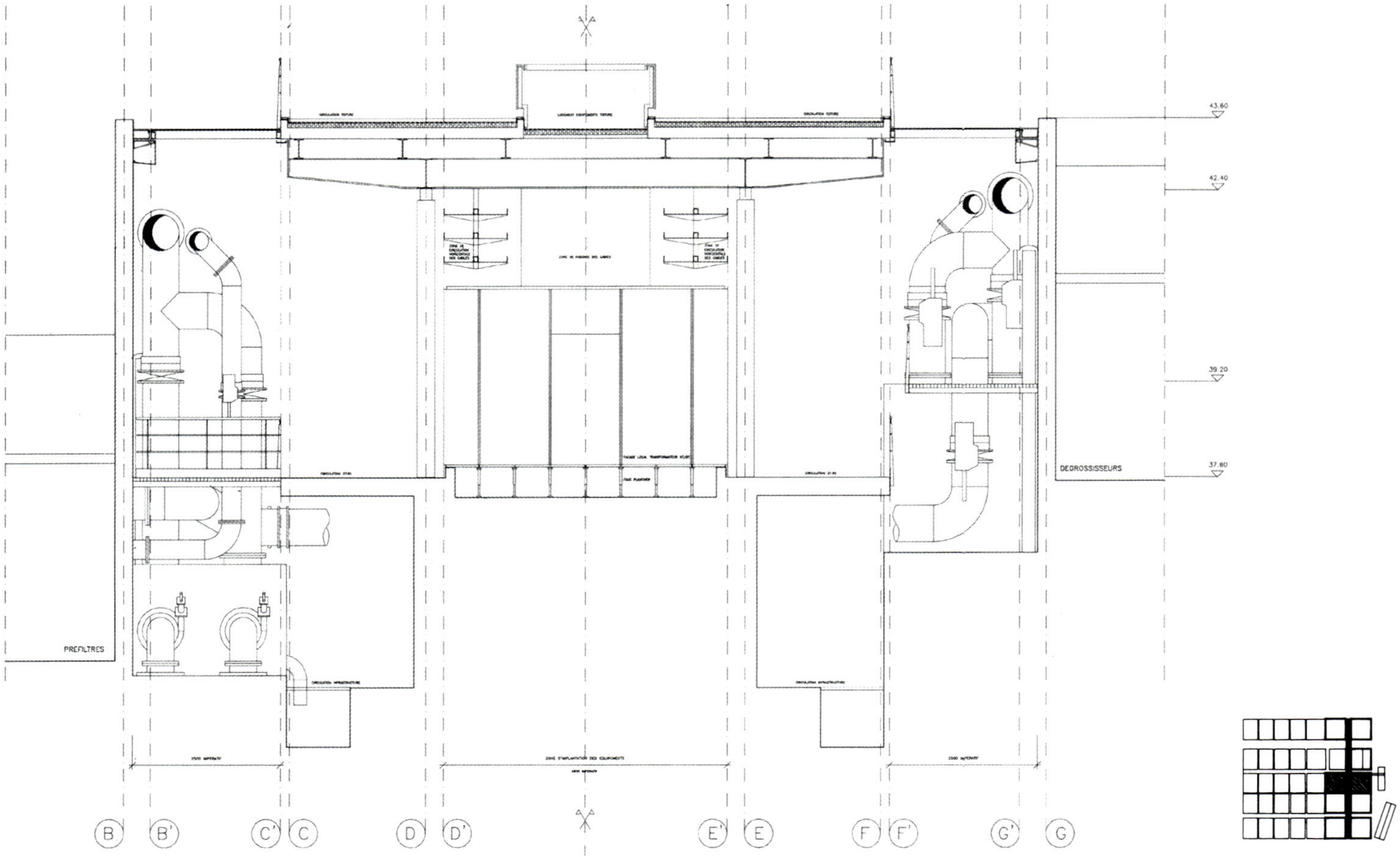

José Gigante / João Álvaro Rocha

Instituto das Comunicações de Portugal (ICP)

Porto, Portugal

Photographs: Luís Ferreira Alves

The site of the Communications Institute stands on the outskirts of the town, due to a fast process of transformation. The weakness of the urban fabric creates difficulties in articulating the characteristic elements of heterogeneous occupations (agriculture, industry, services, lodging). The architectonic disqualification of this area occurs above all from the mischaracterization of the interposed spaces —leftover areas where any design strategy is absent.

The site, near a small farm nucleus, is located between a social housing complex and agricultural fields. The alignment of the street and the choice of a volume similar to those in the neighborhood are devices that, when linked with the reduced dimension of the program, support the relationship between lot and building. The project makes them inseparable: to the performance in the first corresponds the neutrality of the second.

The design of the building results from the re-designing of the ground itself — in the way that resting areas are established in the approach to the building, helping to fix a scale to the whole. The organic relation between building and lot is essentially designed from a concept of mutual assumption, and also by the way that the interior of the building depends on a compromise with the surrounding garden. At the vertical line, a connection is established with the ground via the basement, followed by the elevation volume (work and service areas) and, finally, the upper floors. Along the horizontal line, a hierarchy of spaces seeks optimal exposure to the sun in the work areas.

The exterior facades are also interior — the narrow metallic film that separates interior from exterior is also the element that designs their continuity. The memory of the exterior is thereby present in the interior.

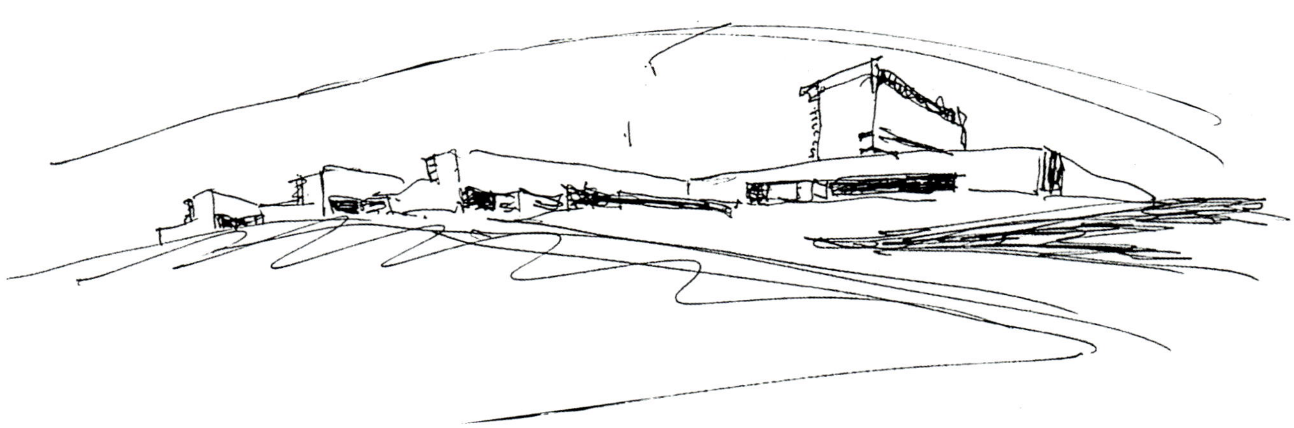

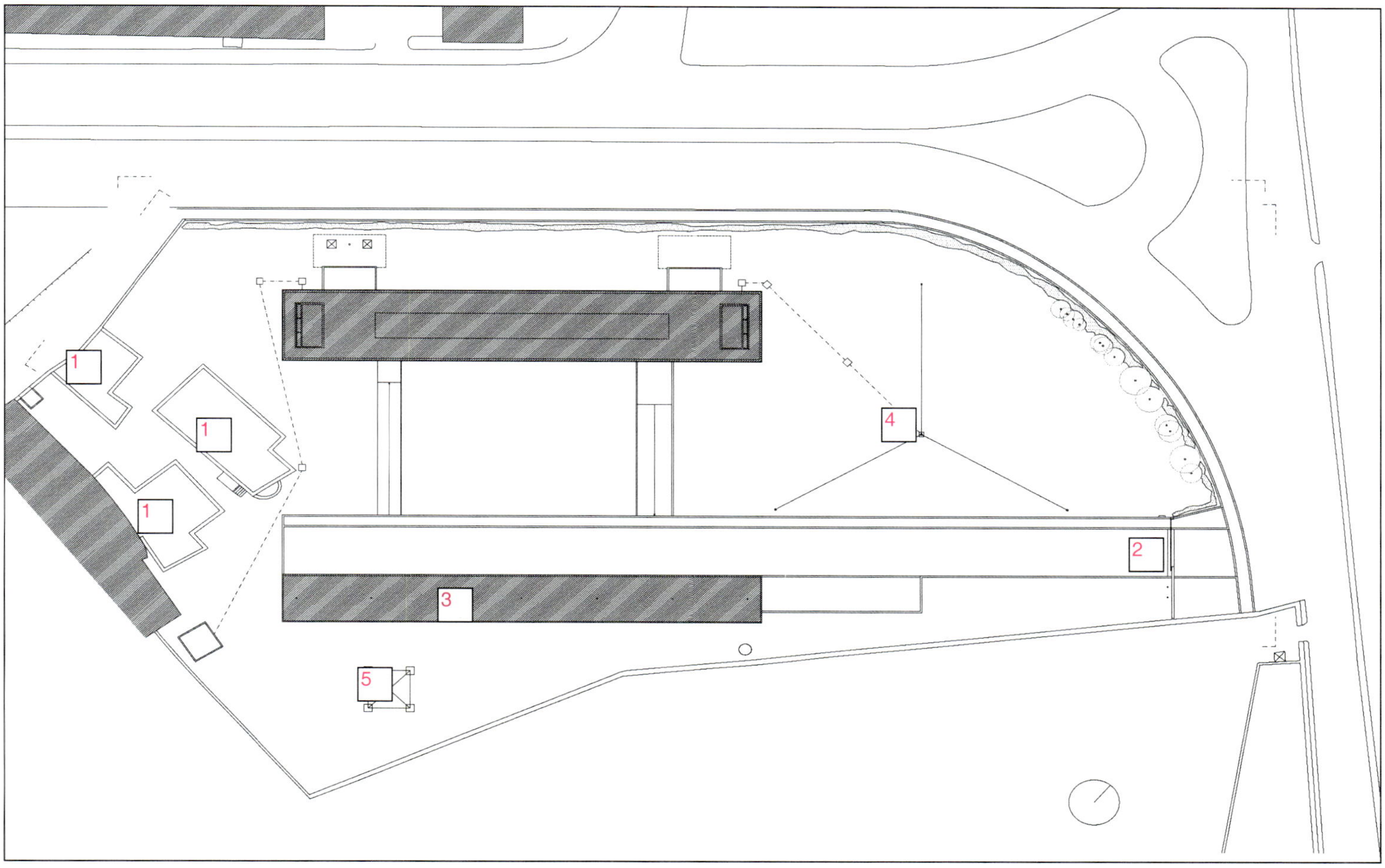

Site plan

1. Ruins
2. Entry
3. Parking
4. Metal tower
5. Self-supported metal tower

The exterior facades are also interior –the narrow metallic film that separates interior from exterior is also the element that designs their continuity. The memory of the exterior is thereby present in the interior.

Cross-section

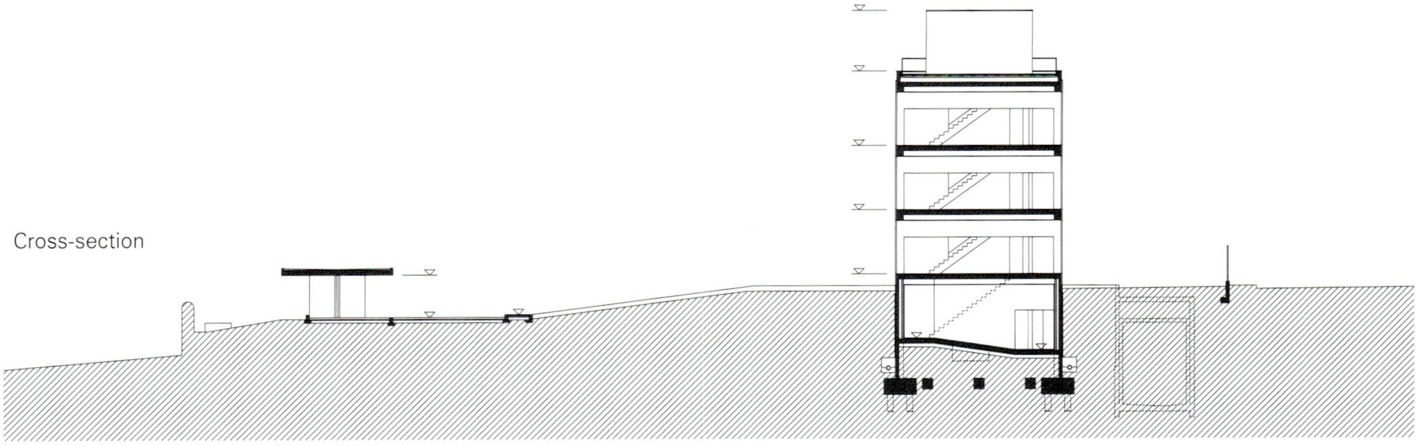

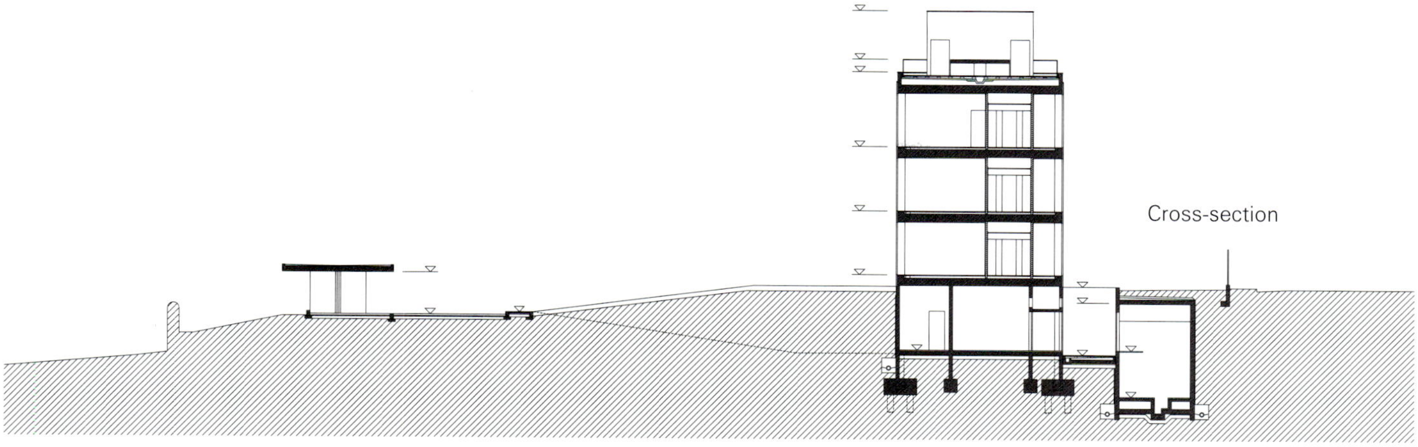

Cross-section

At the vertical line, a connection is established with the ground via the basement, followed by the elevation volume (work and service areas) and, finally, the upper floors. Along the horizontal line, a hierarchy of spaces seeks optimal exposure to the sun in the work areas.

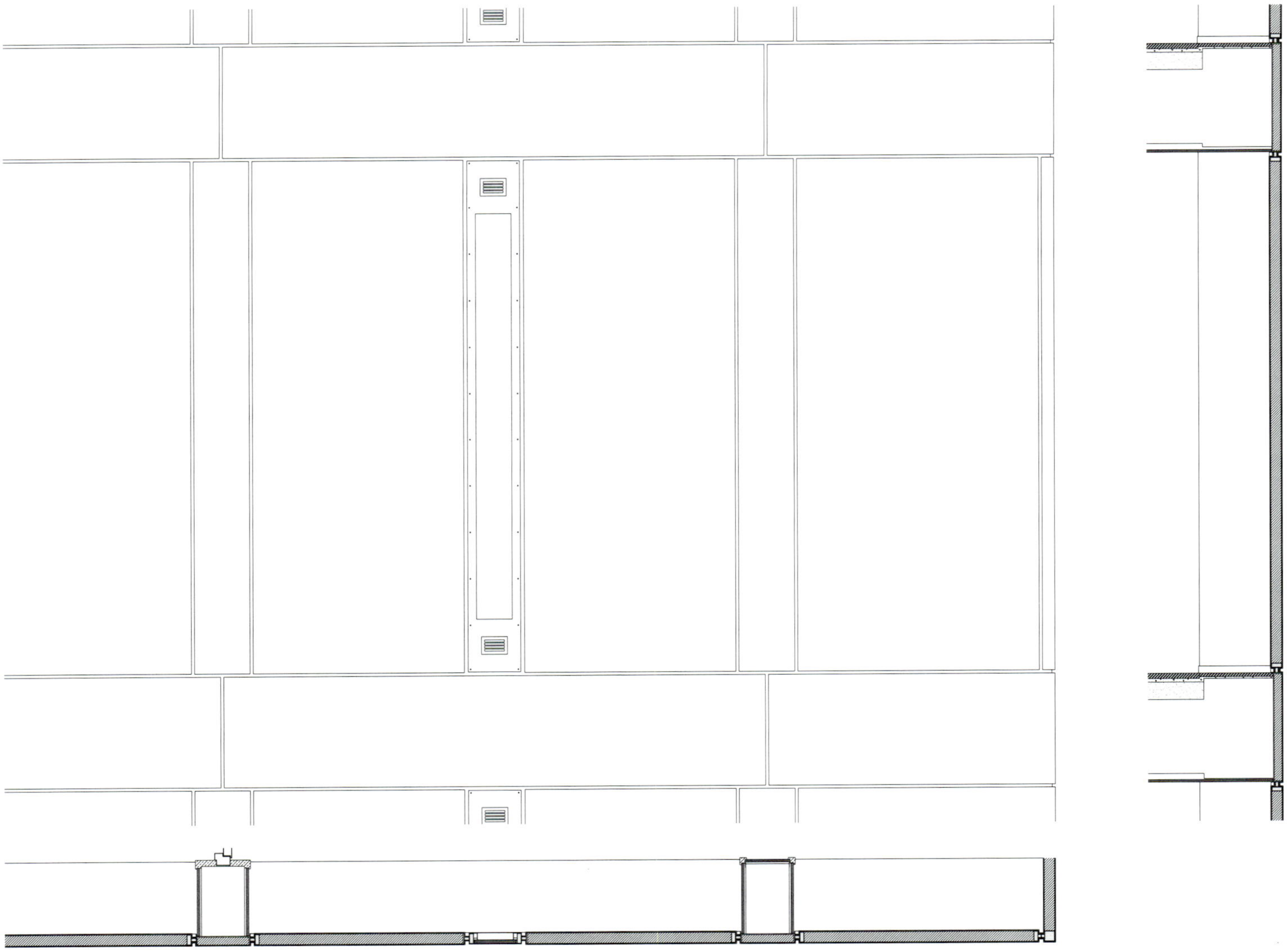

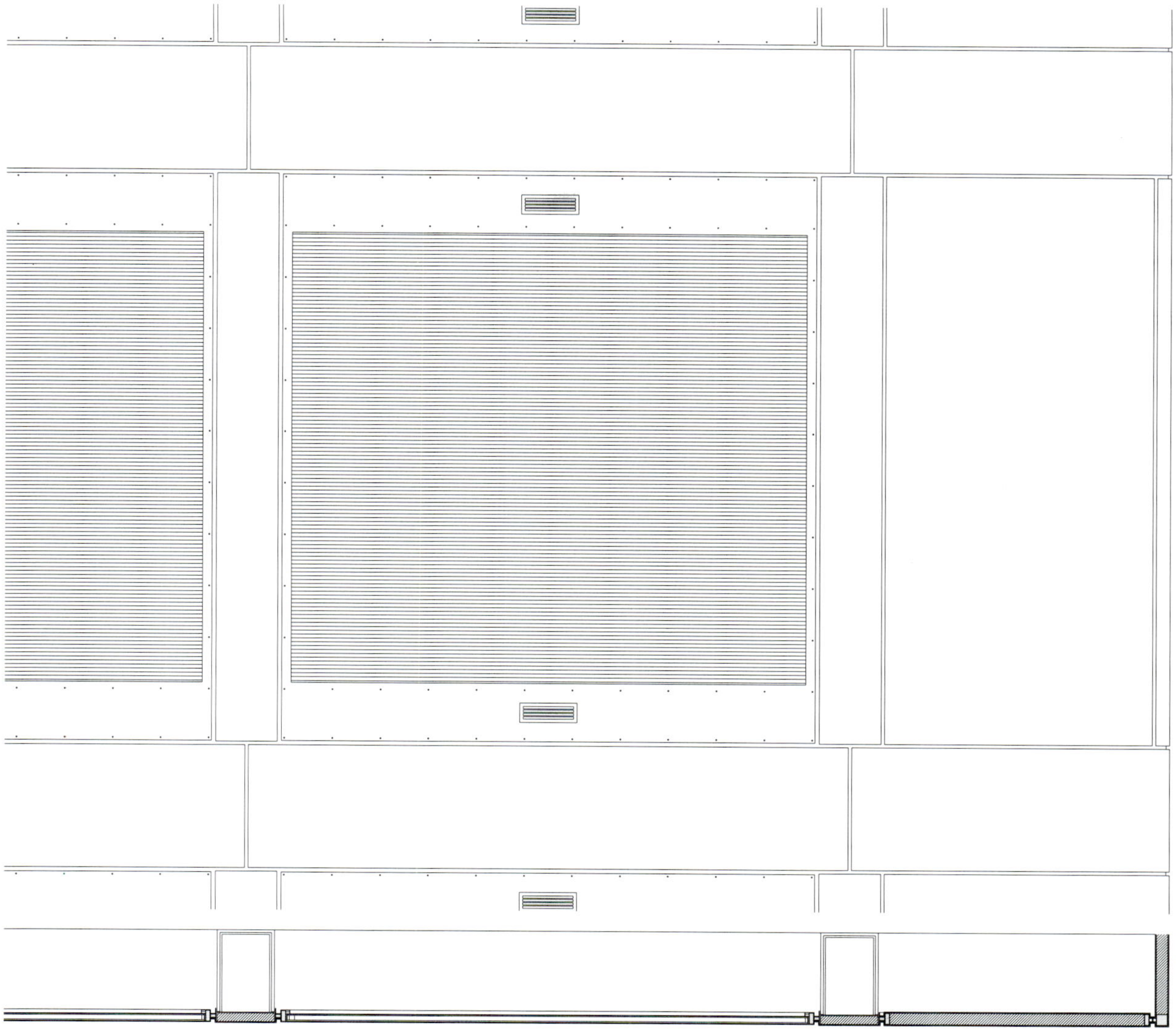

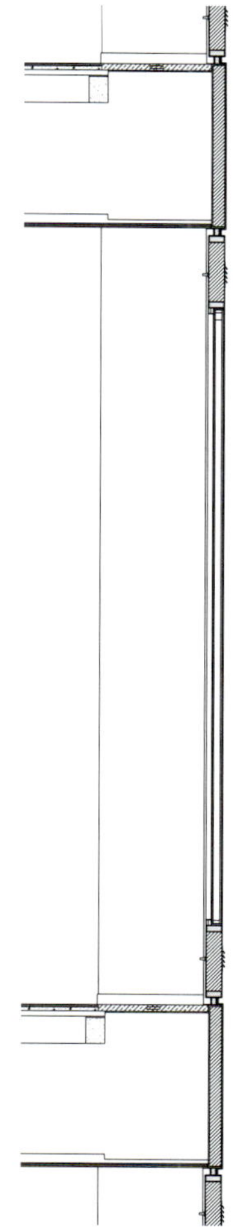

Window details

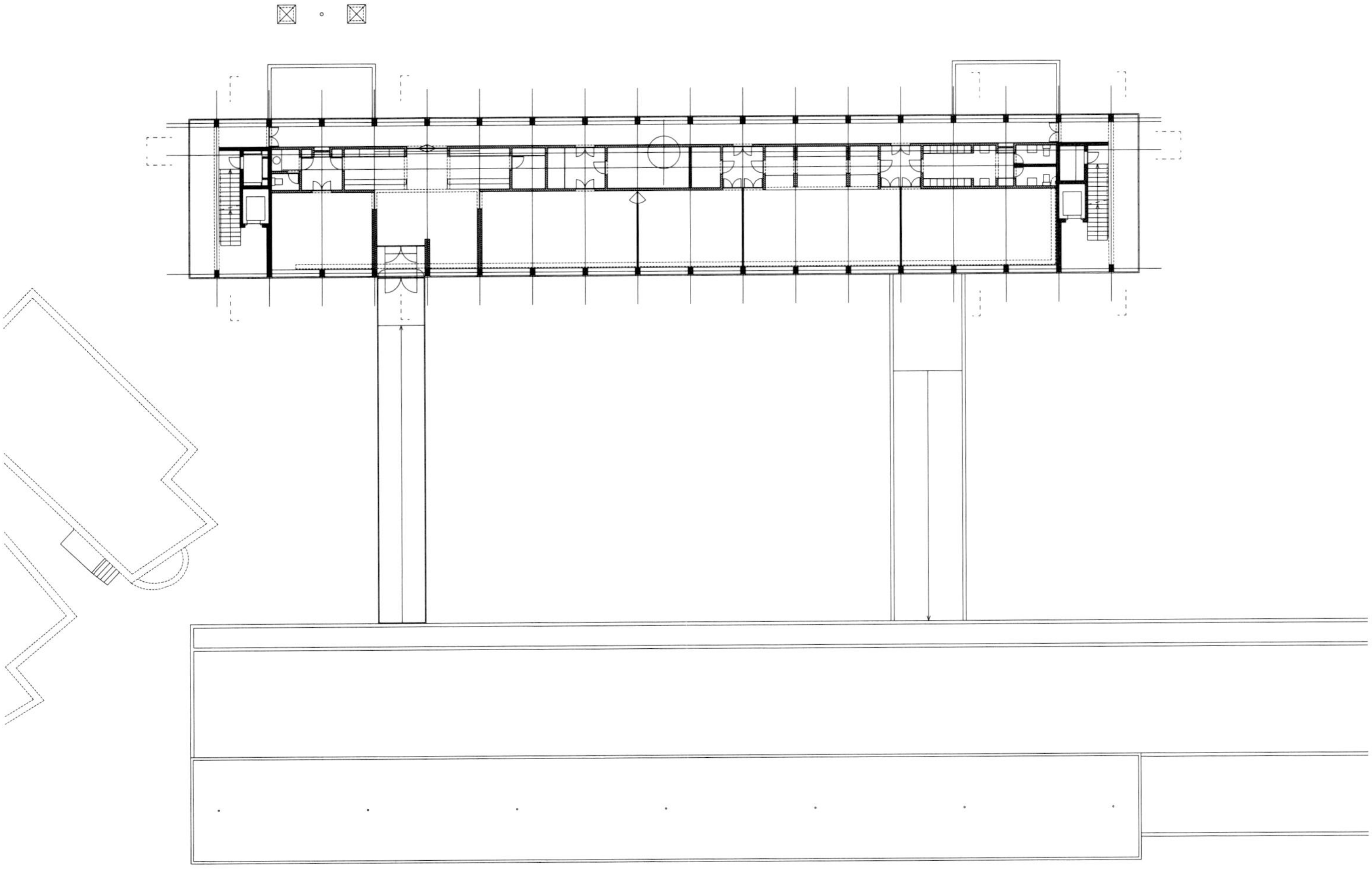

Ground floor plan

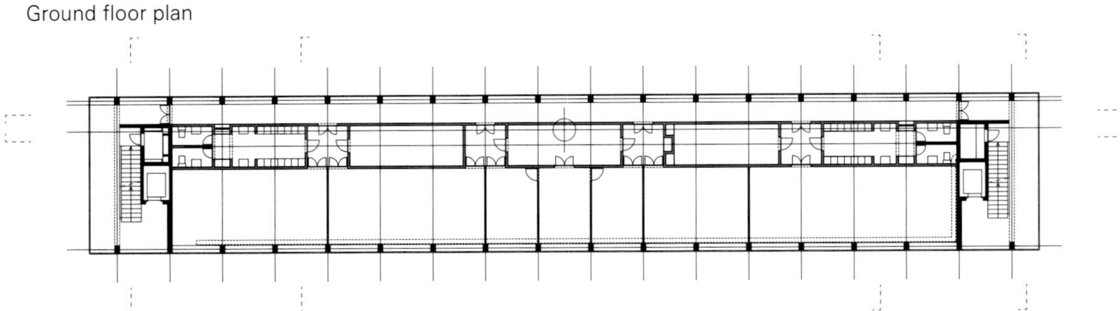

First floor plan

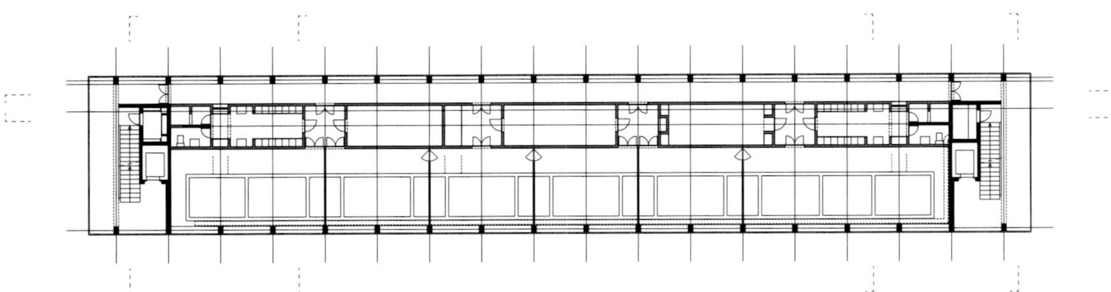

Second floor plan

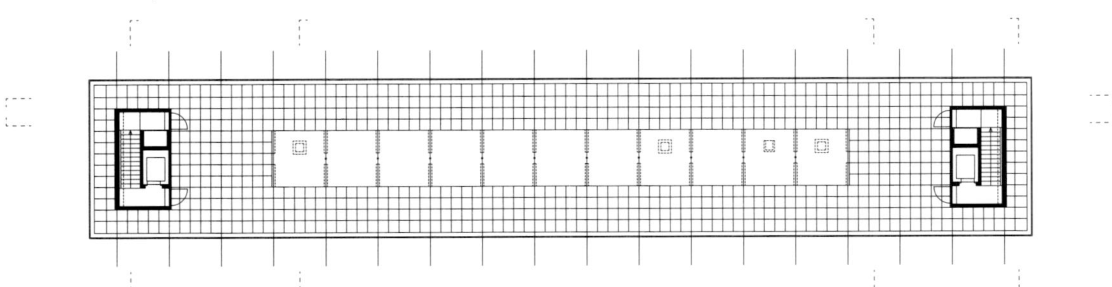

Roof floor plan

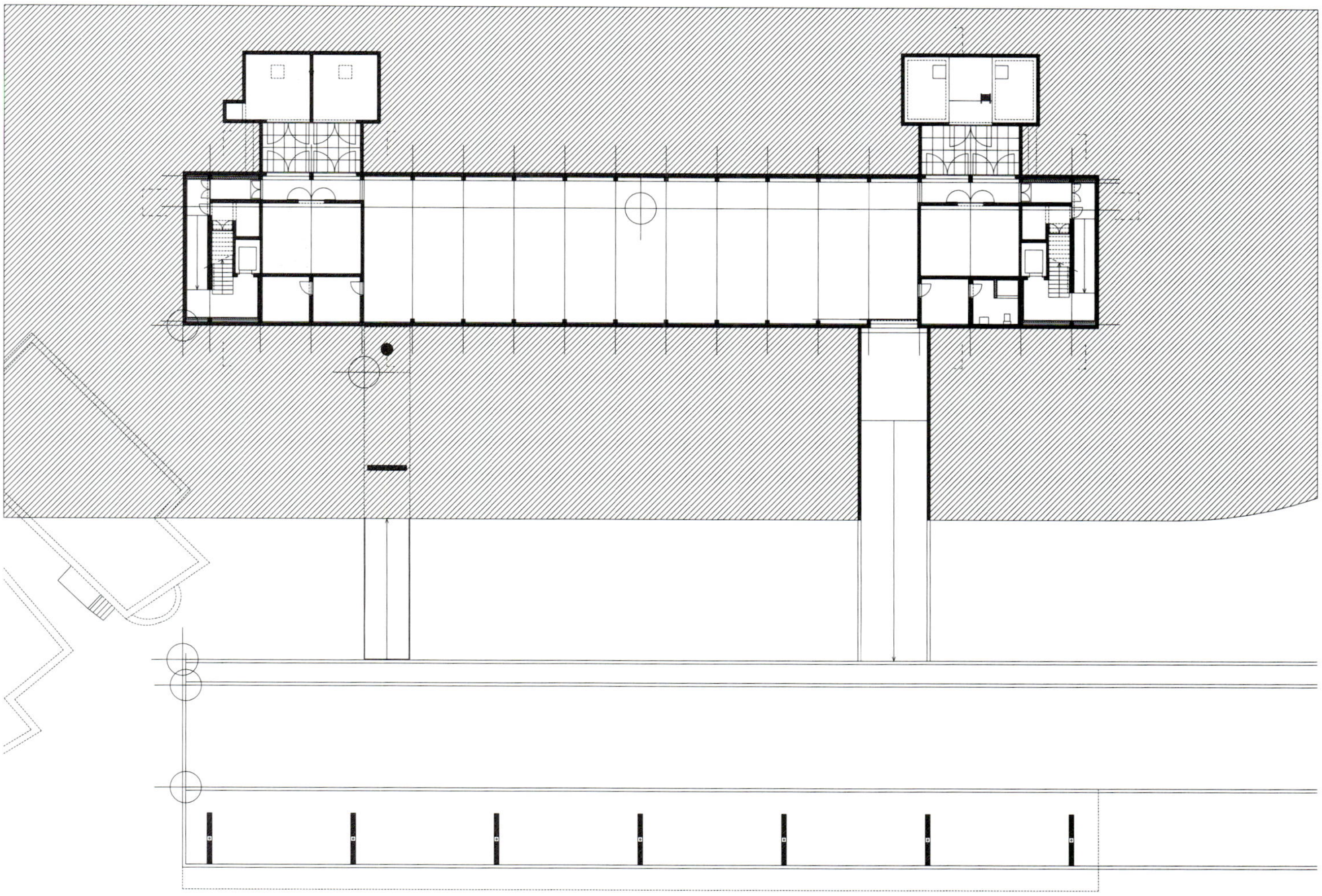

Basement floor plan

Krueck & Sexton Architects

Phillips Plastics Corporation

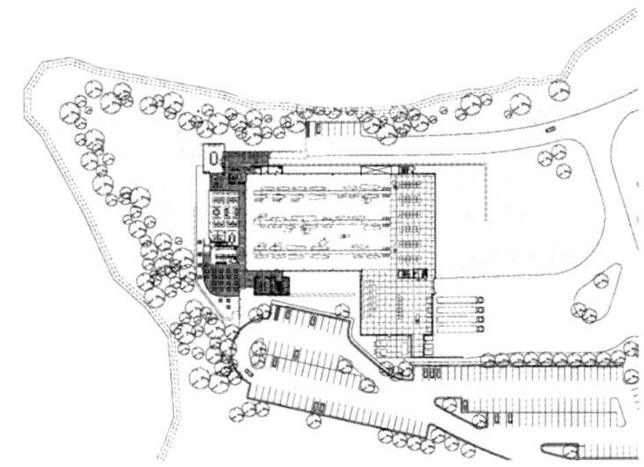

Wisconsin, USA

Photographs: Mariusz Mizera

Designed for an industry leader in plastic molding technology, this 85,000 sq ft custom manufacturing facility advanced the form of the industrial workplace. Located in rural Wisconsin at the confluence of the Elk River and Elk Lake, the building is sited to minimize the removal of trees, while taking maximum advantage of exceptional views in several directions. Because of an extensive use of glass, all employees can enjoy these views from every location in the facility.

Given its northern location, careful consideration is given to the colour of materials and the quality of light entering the workspaces, especially during the winter season. Of equal concern is the quantity of light, which is filtered through the use of shading overhangs, fritted glass coatings and vegetation. These strategies eliminate fatigue-producing glare and reduce energy consumption, while filling the interior with an abundance of natural light.

The building is composed of intersecting glass, aluminium and native Mankato-Kasota stone volumes, arranged to dissolve the distinction between the manufacturing bays and the smaller scale office and common areas. The largest volume, a 220-foot long by 130-foot wide, 35-foot high glass manufacturing shed, is built around gantry cranes facilitating rapid reconfiguration of the three production lines. An acoustically rated glass wall, the only element between manufacturing and administration, preserves internal vistas, connectivity and openness.

Amenity spaces which are shared by all are located at the perimeter, transitioning the manufacturing shed to the landscape. The weaving of materials such as stone, into both exterior and interior further encourages the movement of employees into the surrounding wooded site.

At the request of Phillips, the project team utilized a non-traditional approach to the design process. Executed by a design/build delivery system, it involved intense interaction and collaboration between owner, architect, contractor and subtrades, resulting in exceptional craftsmanship and quality with substantial savings in cost and time.

FN7000

phillips
phillips
phillips

Natural light is filtered the use of shading overhangs, fritted glass coatings and vegetation. These strategies eliminate fatigue-producing glare and reduce energy consumption, while filling the interior with an abundance of natural light.

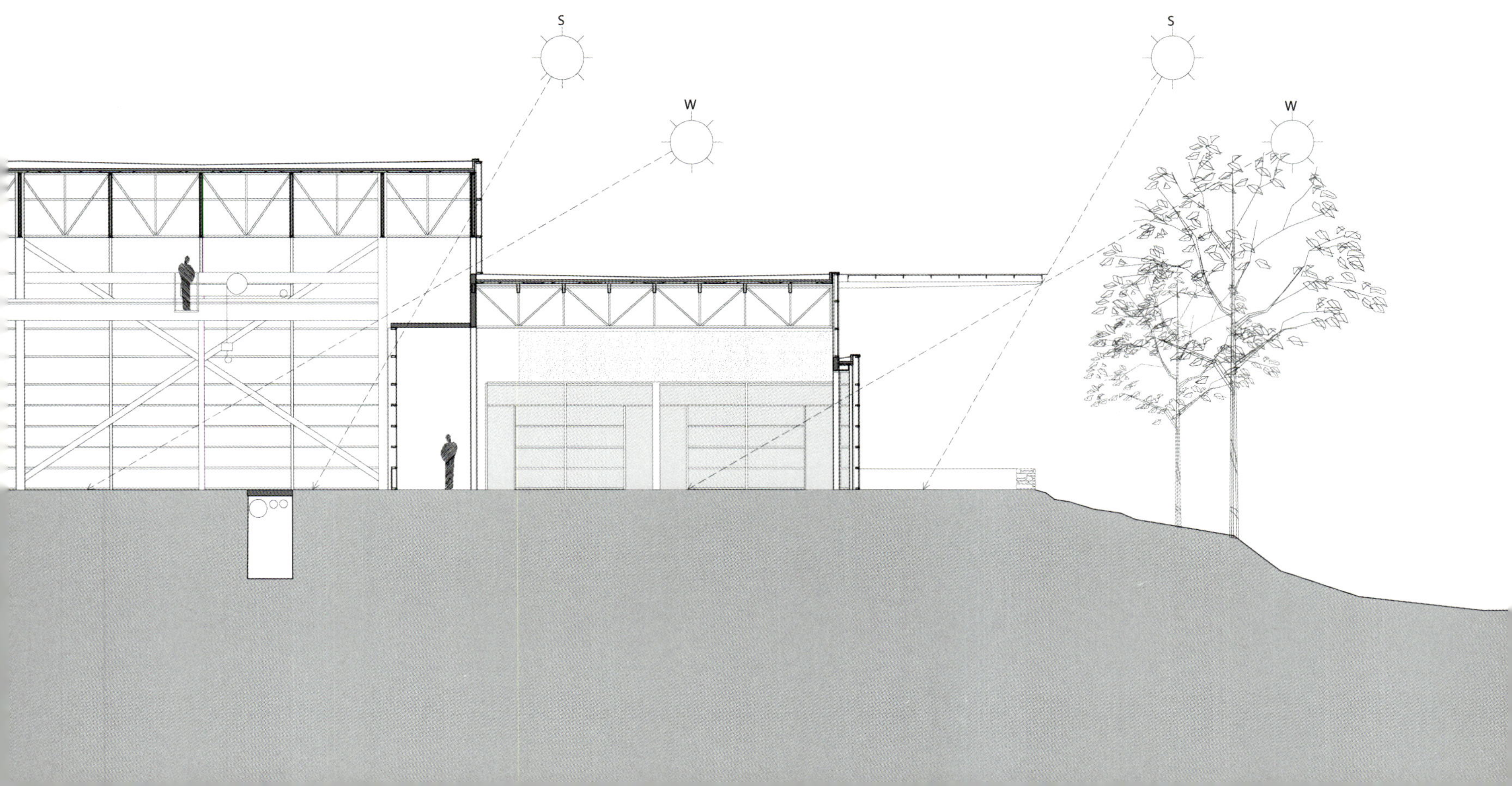

NL Architects

Wos 8

Utrecht, The Netherlands

Photographs: NL Architects

WOS 8 is a Heat Transfer Station (Warmte Overdracht Station) for the latest extension of Utrecht, the new city of Leidsche Rijn. The water cooling the turbine fans of the large UNA power plant, about a kilometer away contains enough energy to supply all new dwellings with heat and warm water. Before, this energy was dumped as waste heat into the Amsterdam-Rhine canal. Now, it will be recycled and put to use; a zero emission program. This building is the node where the loop from the power plant transfers its energy to several distribution loops serving the future housing areas. In total WOS 8 will service 11,000 dwellings. The employed technology consists of a central heating system on the scale of an entire city.

The size is determined by the minimal functional dimensions of the system of pipes in the interior and forces of realty economics to keep it as small as possible. Architecture reduced to a skin-deep zone. Due to the internal organization, the south side didn't require the full height. Here, some manipulations of the volume were imaginable.

In the future, when the city has expanded into these farmlands, this structure will be situated in the middle of the contemporary idyll of a new suburbia, with a public access road directly abutting the east façade of the building. From then on WOS 8 will be a tactile part of the public domain.

A polyurethane membrane makes for seamless architecture. The material was originally developed for roofs: strong, flexible, waterproof, durable, attractive and chemically inert (no pollution of earth and ground water). It is easily applicable by spray gun or paint roller. Building parts that might refer to scale or size, like doors, disappear.

Usually these complexes are the blind spots of urbanism, as well as easy targets for vandalism. Not in this case. The facade is programmed with a number of elements meant to make the building attractive; WOS 8 is a public square wrapped around a box. And aims to become part of the youth culture that usually is the biggest threat for this kind of building.

The climbing grips spell "the Blind Facade" in Braille. A "door-scope" is placed the wrong way around in the main door. Normally used to see who is at the door, here it reveals the interior. This area receives an average of 134 rainy days a year, a fact which inspired the reinvention of spouts and cisterns, while the polyurethane skin lets the rain play sculptural tricks with the building.

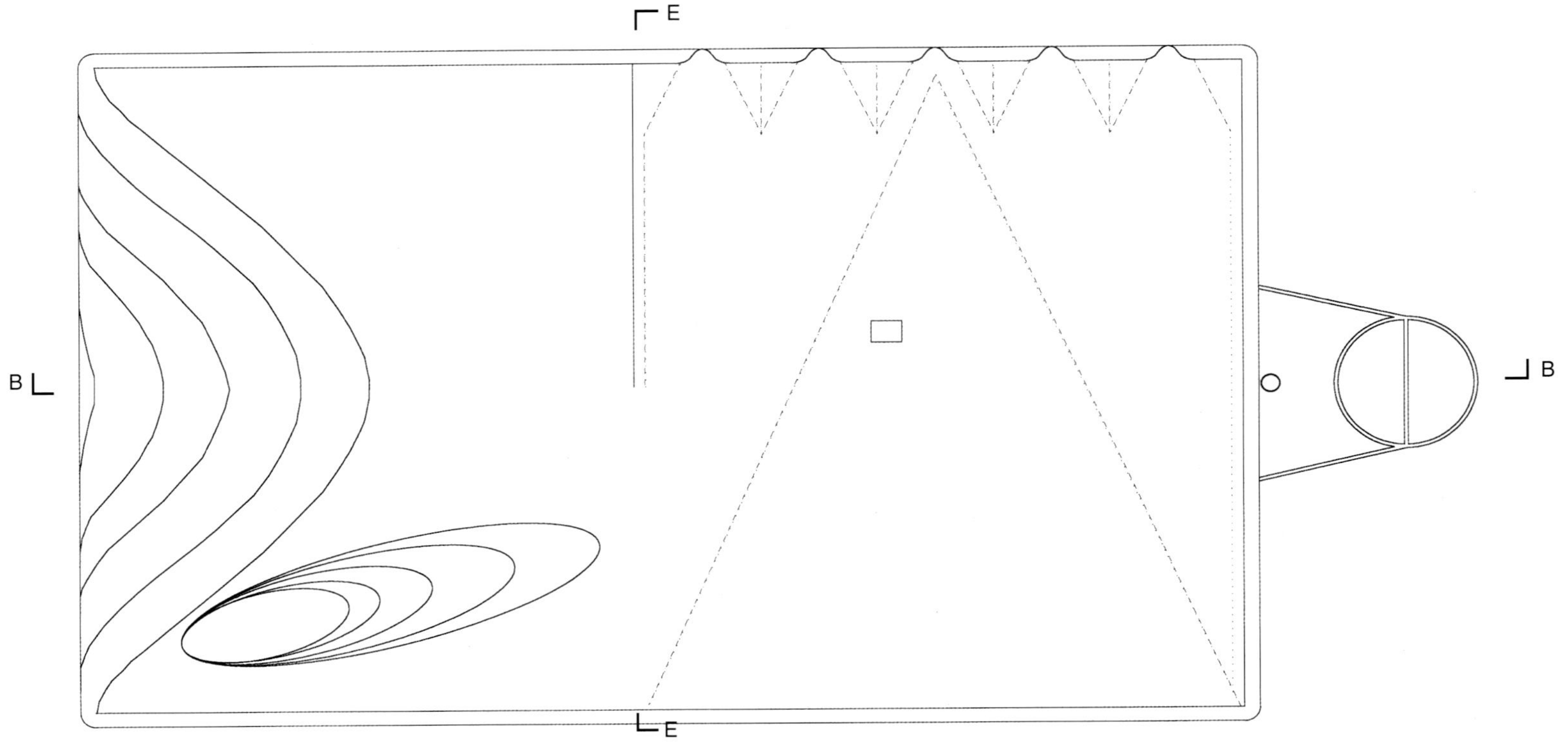

Roof plan

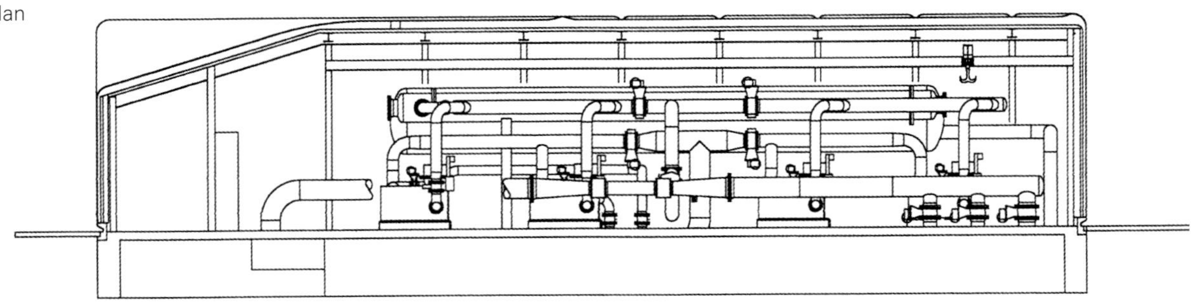

Cross-section B-B′

Floor plan

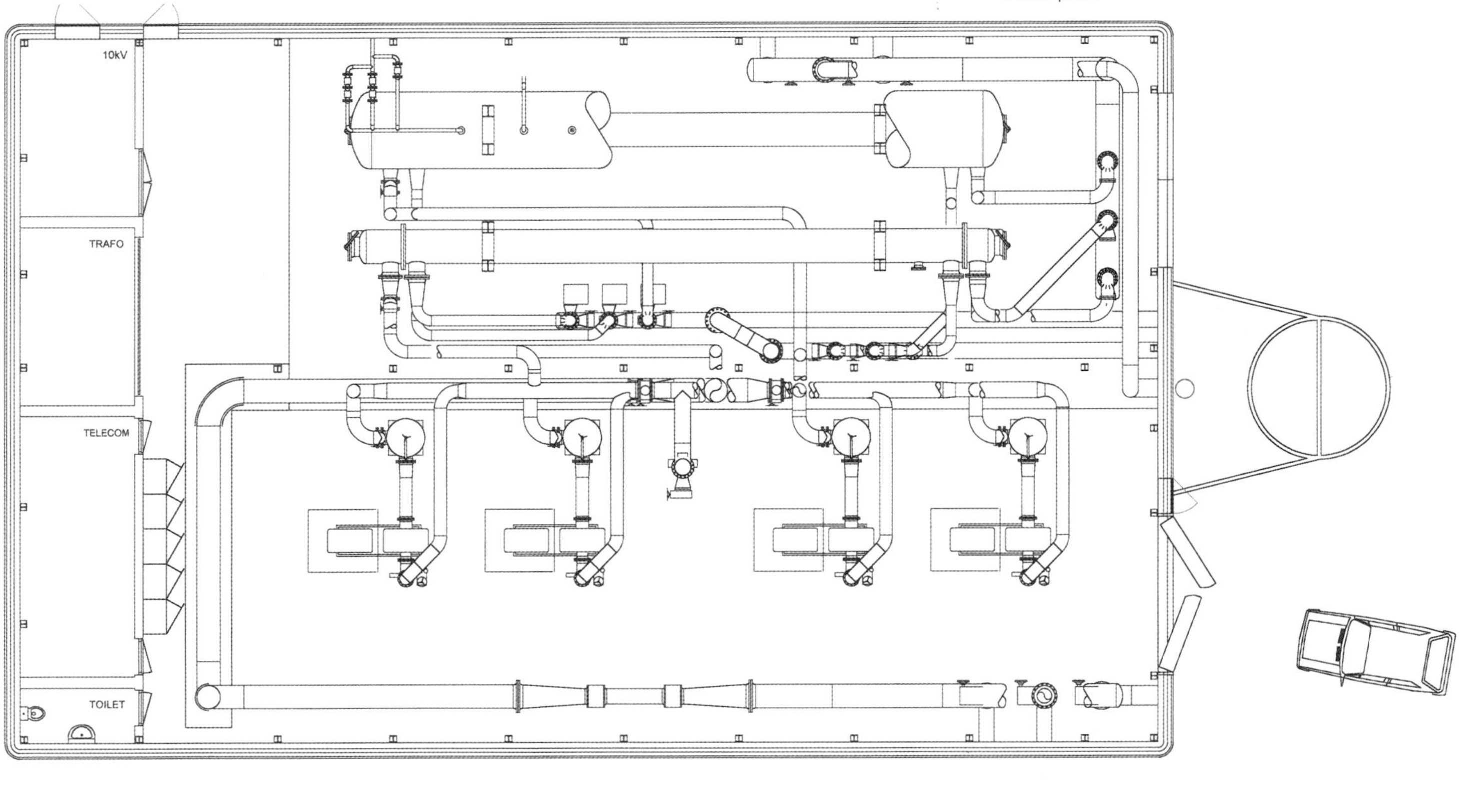

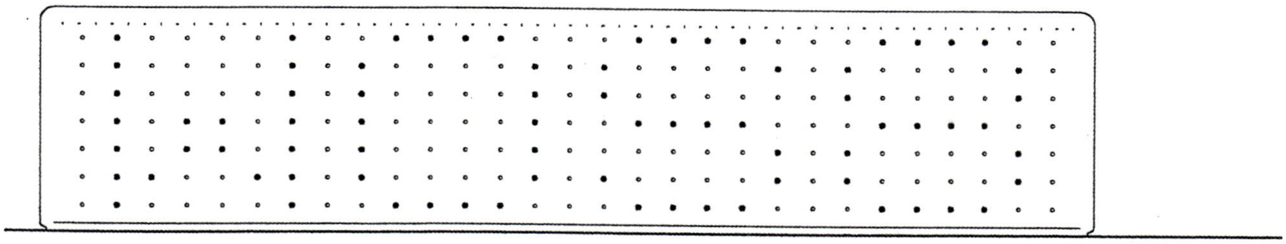

East elevation

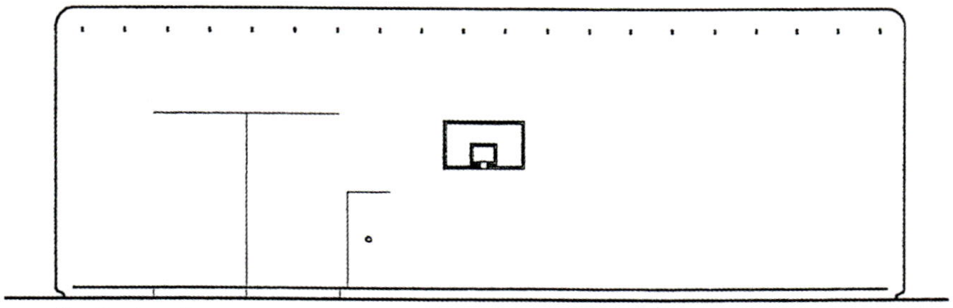

North elevation

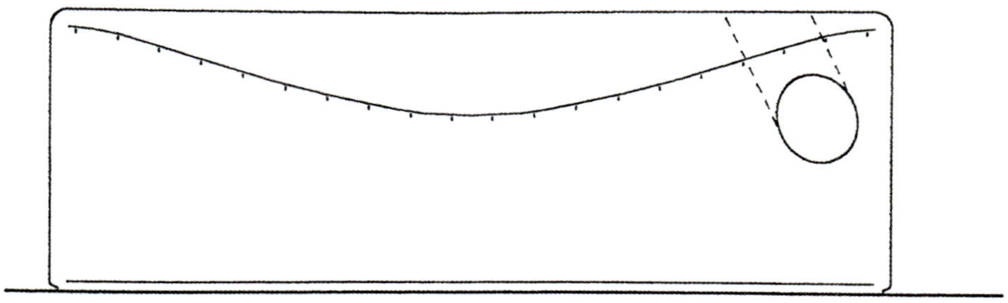

South elevation

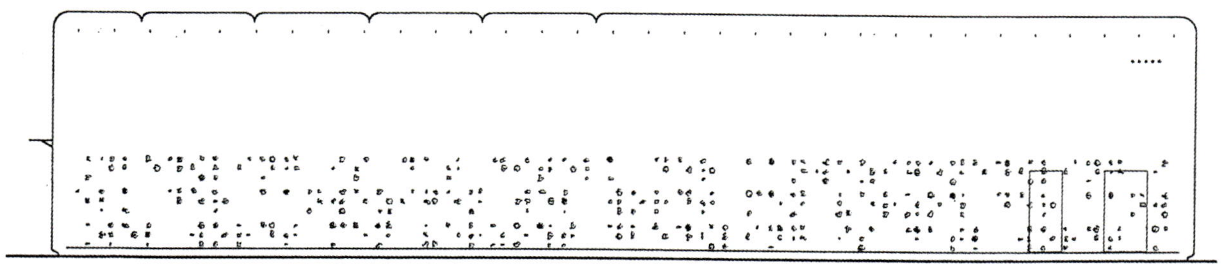

West elevation

RoTo Architects

Warehouse C

Nagasaki, Japan

Photographs: Katsuhisa Kida

Japan's Nagasaki Prefecture has recently created a new landfill pier, approximately 800 feet in length and 400 feet in width, that projects into Nagasaki Harbor. The entire pier will be mixed-use, with a ferry terminal, large retail structures and working warehouses. The city, outlying mountains, and bay sorround the site, making it highly visible from nearly every location. Around the harbor, the site can be viewed by the many passenger ferries and ocean liners that link Japan to the world. With this in mind, the architects proposed a rooftop public garden space in addition to the 14 meter high, two level warehouse that had been commissioned to design.

Warehouse C is 200 meters long, 25 meters wide and an average of 22 meters high, dimensions comparable to a super tanker. It is organized in three essential parts: a two level private warehouse on the bottom, a public garden on its roof and a twenty meter diameter spherical exhibition hall. The roof scape, a new elevated ground floor plane, provides a public link between the new ferry terminal and retail spaces on the new wharf and the center of the city. The garden will be a present day version of the traditional "dry" gardens of Kyoto.

The overall building is a composite of three construction systems: the warehouse is concrete; the south elevation wall system and half of the roof enclosure are steel plate and the other half of the garden is enclosed with fabric on the north side. The spherical exhibition hall is insulated steel plate, 20 meters in diameter, which is normally used for petroleum storage on the huge tanker ships built in Nagasaki. The steel shields the building from the prevailing winds of the annual monsoons.

The steel structure, plate wall, roof systems and the steel sphere were manufactured by shipbuilders across the harbor and brought to site by barge.

The rooftop and the walls are shaped by a series of non-parallel frames that are divided into four primary zones. Each zone defines a different type of experience on the roof by changing the scale and degree of the enclosure of the space. The form that results from the twisting and undulating geometry of the fabric and steel planar systems is a variation on the conventions of the complex planar geometries of shipbuilding. The shipbuilders engineers said these configurations and the construction logic to attain them was conventional, while the building engineers and contractors claimed that the opposite was true.

JA長

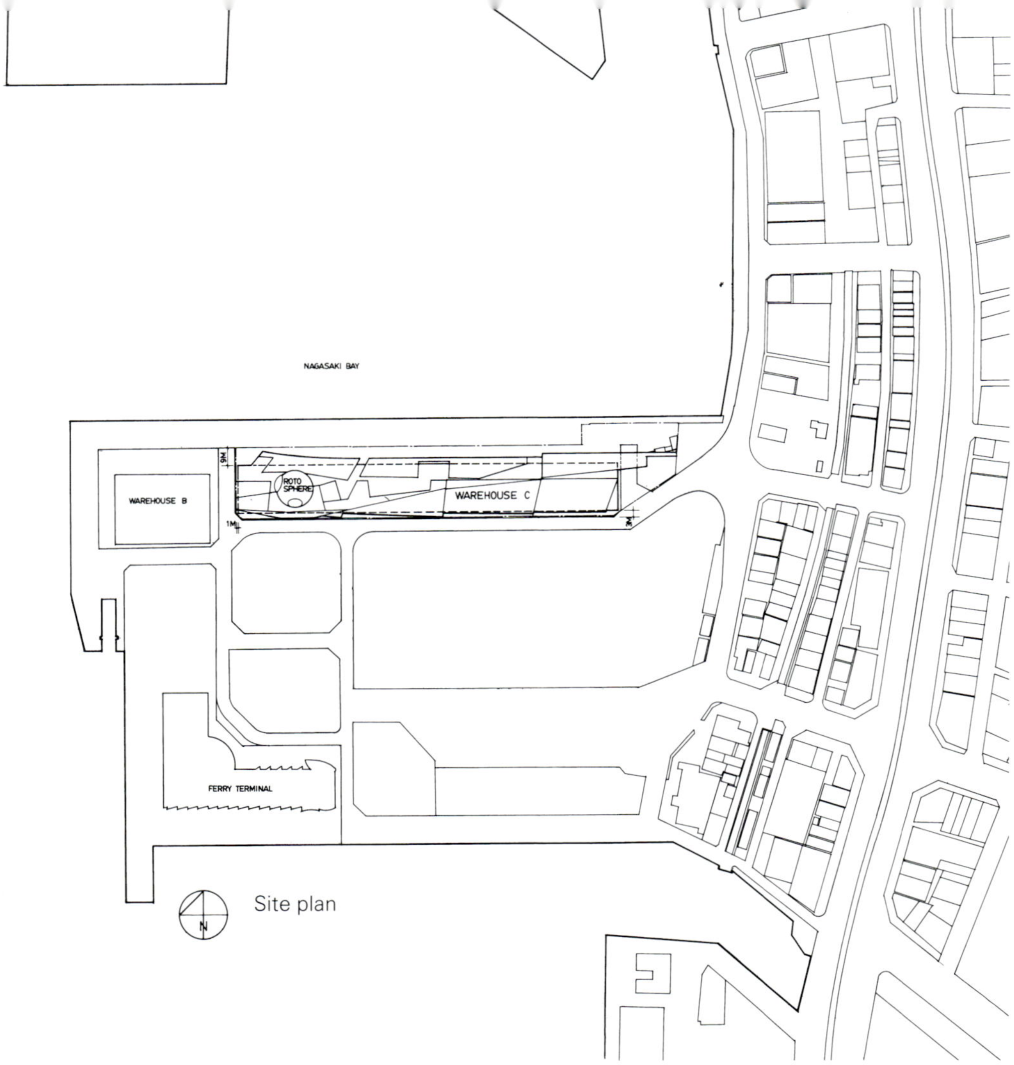

Site plan

The design drew on the techniques used in shipbuilding. The spherical exhibition hall is insulated steel plate, a material normally used for petroleum storage on tanker ships, but which here shields the building from monsoon winds.

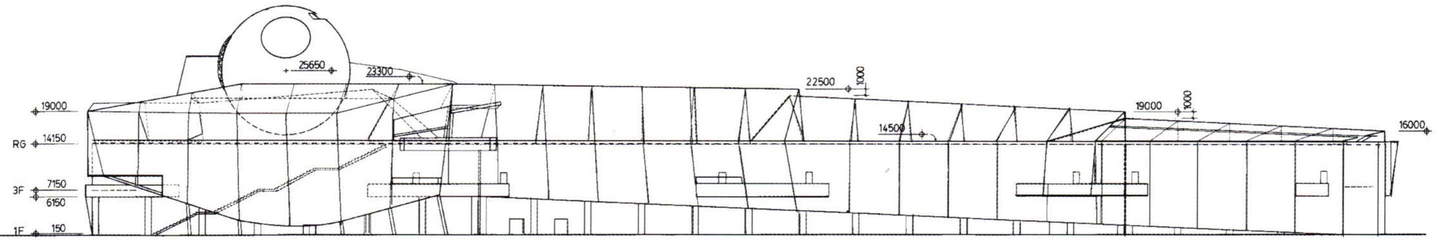

North elevation

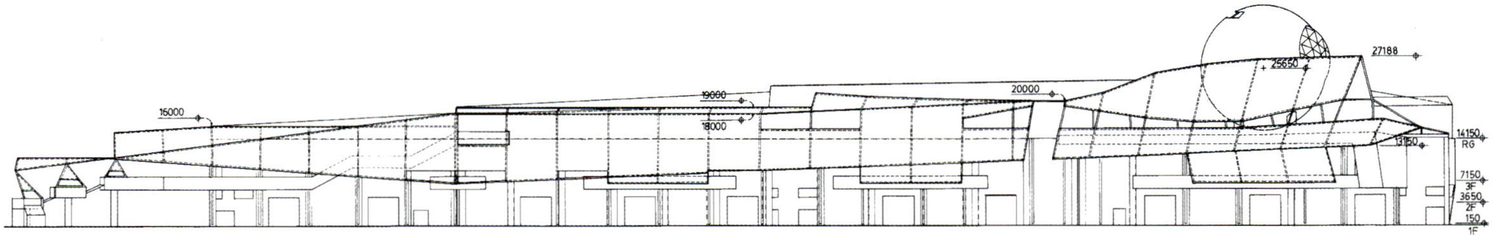

South elevation

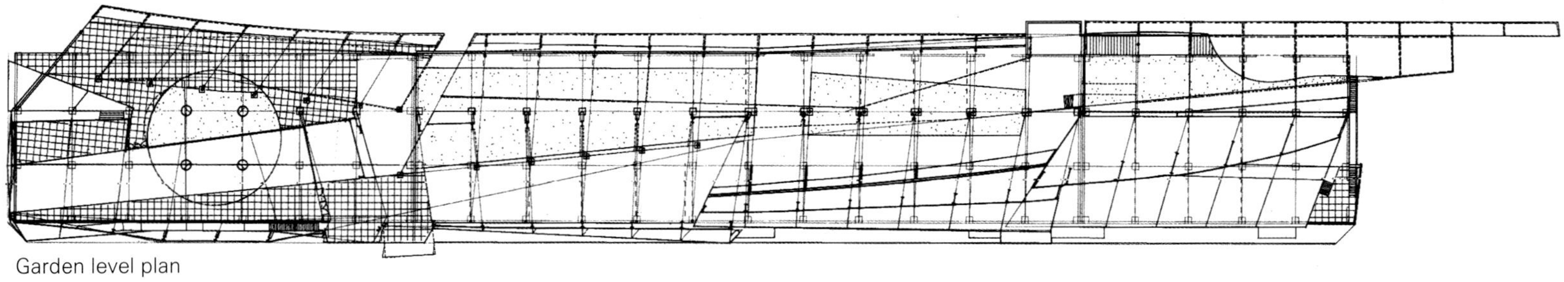

Garden level plan

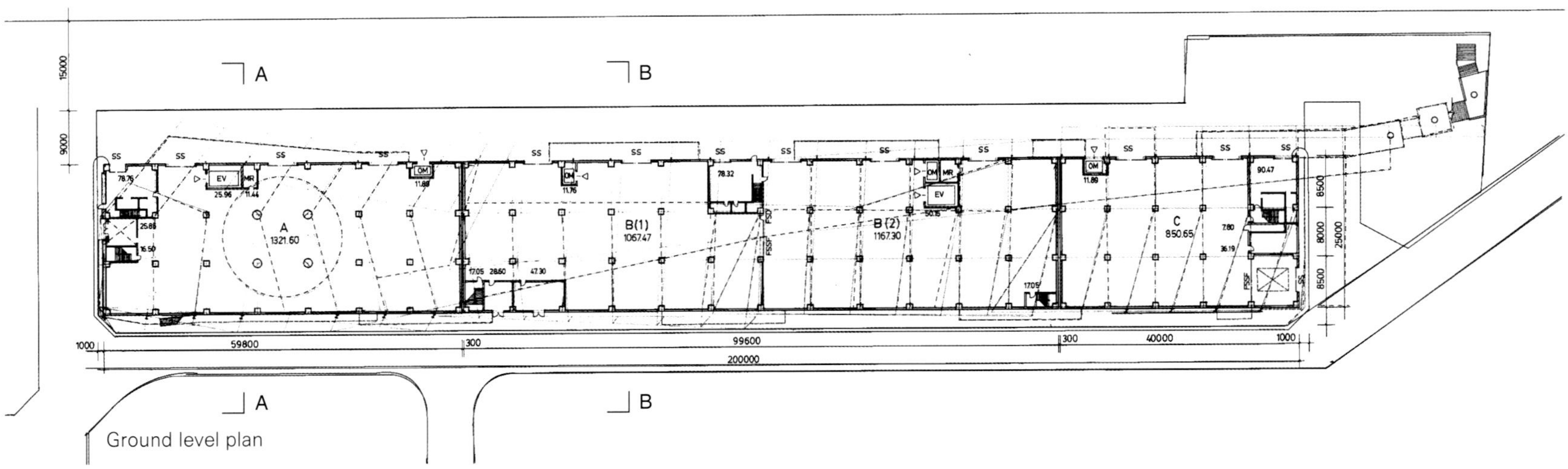

Ground level plan

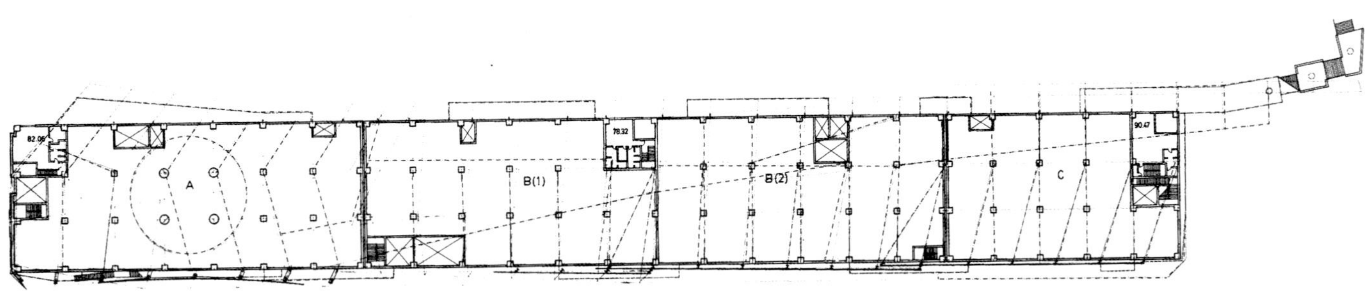

Second level plan

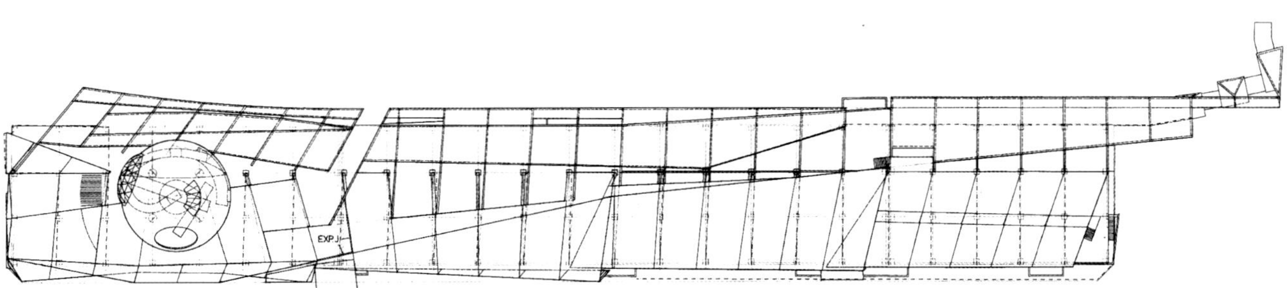

Roof plan

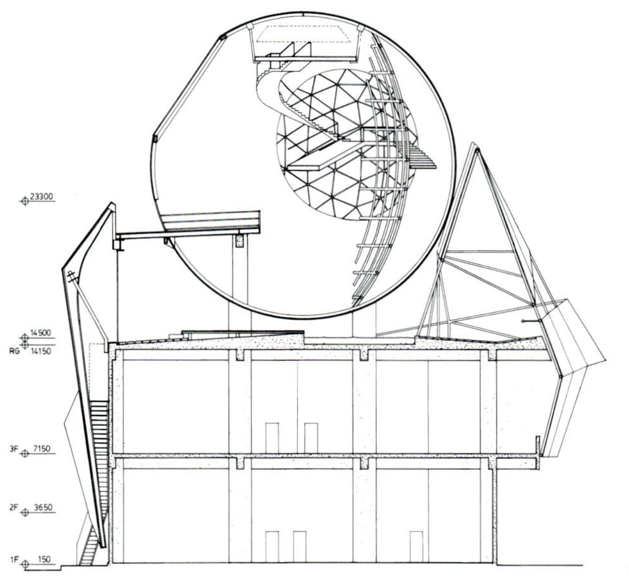

Section A

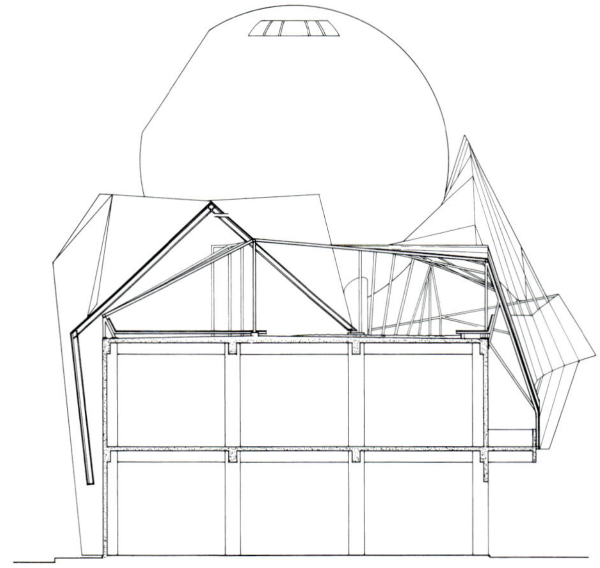

Section B

The rooftop and the walls are shaped by a series of non-parallel frames divided into four primary zones. The form that results from the twisting and undulating geometry of the fabric and steel planar systems is a variation on the conventions of the complex planar geometries of shipbuilding.

Tectoniques

Centre Loguistique

Lyon, France

Photographs: Christian Michael

The site for this project is "Les Subsistances", a former military installation, which was occupied by a heterogeneous building complex devoid of all architectural or urban coherence which had fallen into a state of neglect. The architectural intervention consisted of reorganizing and completing this logistic platform with supplementary storage areas and changing rooms.

In clarifying the program, certain elements were demolished, while others were renovated. The external storage and parking areas had to be restructured and a new complementary wing was built.

The building that was created is intended for storing a vast range of materials: barriers, podiums, polling booths, electoral hoardings, flags, and sundry materials and furniture in transit. The exterior skin features three general groupings of material. The first group, which could be defined as mineral, is represented by prefabricated, stained concrete panels (of which the largest single piece occupies 20 sqm); a translucent category comprised of panels of tinted glass; and a composite approach, which unites a layer of metal panels, a layer of insulation, a panel of transparent polyester and an outer layer of black-tinted, perforated metal.

From the viewpoint of visual permeability, the last sandwich is intercalated between the first two, suggesting a mysterious presence of the background thought the perforations in the foreground.

The end result is a covered, hollow-bodied polygon formed by a concrete skeleton that releases the useful space from all load bearing elements and whose dimensions are established according to the movement of fork-lift trucks.

On the other hand, the sectors that have been rehabilitated (storage and changing rooms) restore the original form and language, particularly characterized by an enormous expanse of saw-tooth shaped roofs.

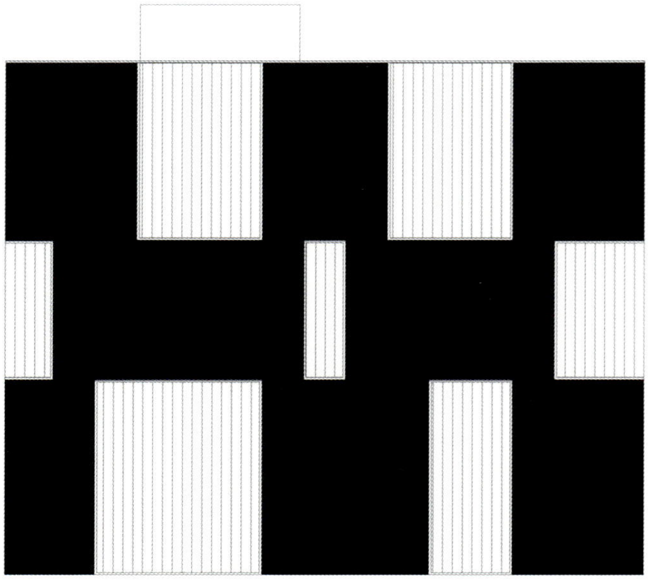

The renovated sectors (storage and changing rooms) restore the original form and language of this former military installation, particularly characteri-zed by an enormus expanse of saw=tooth shaped roofs.

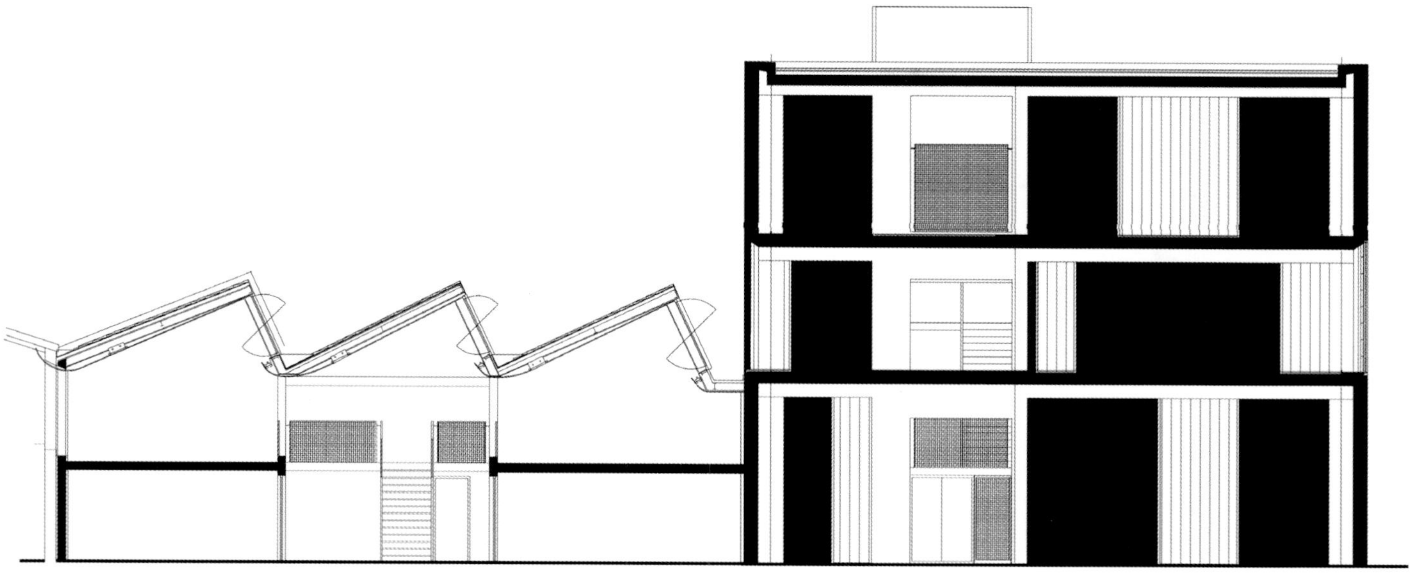

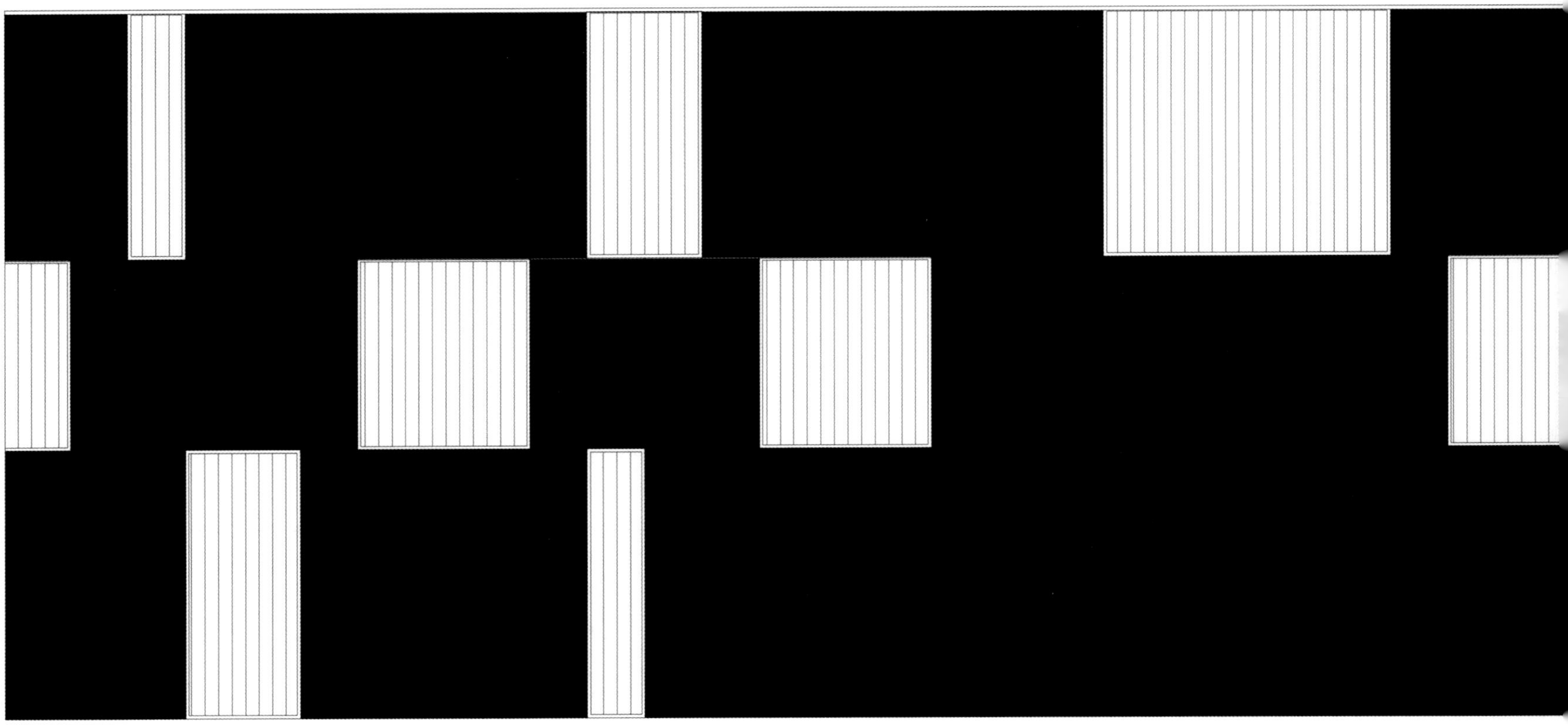

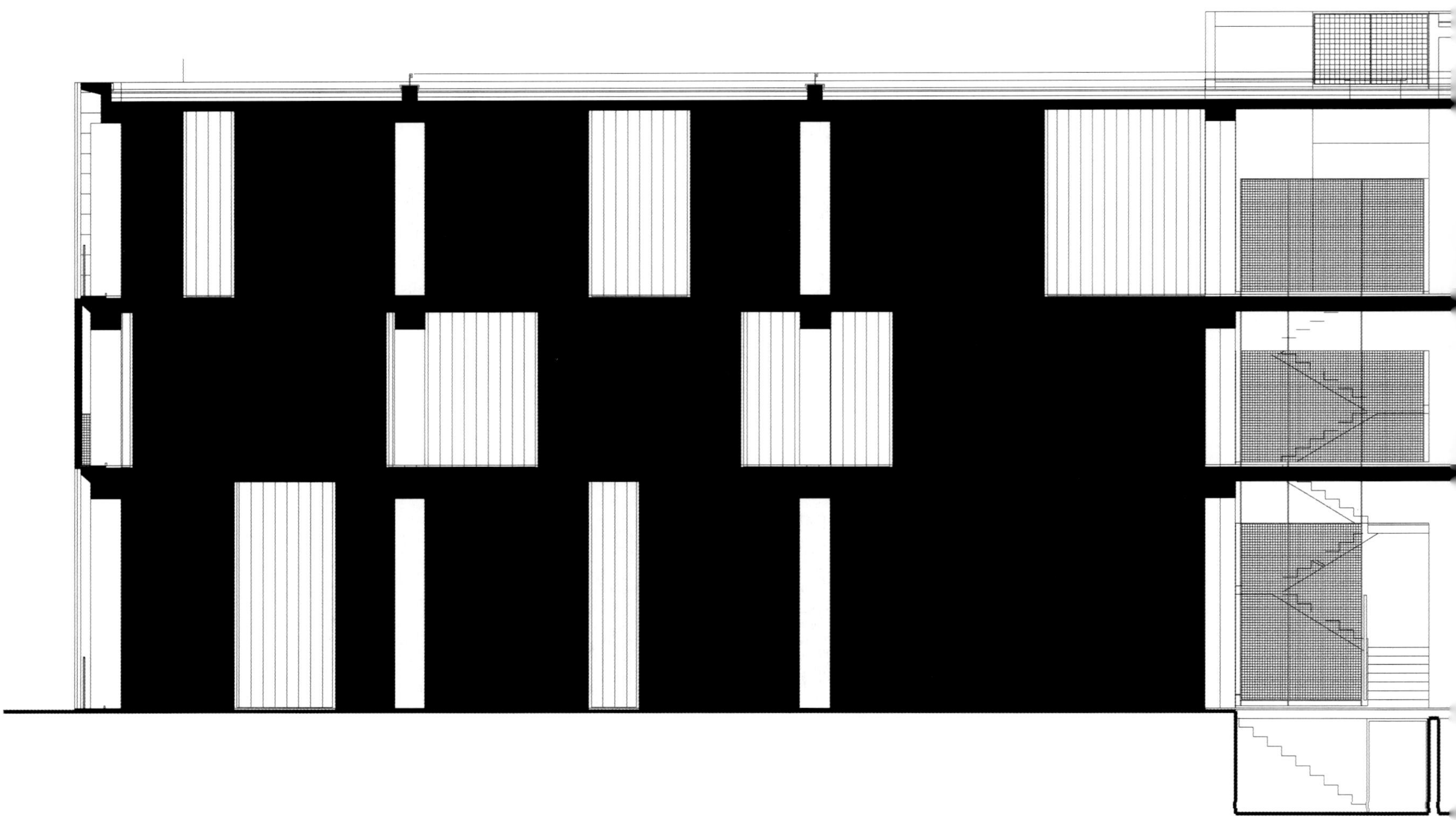

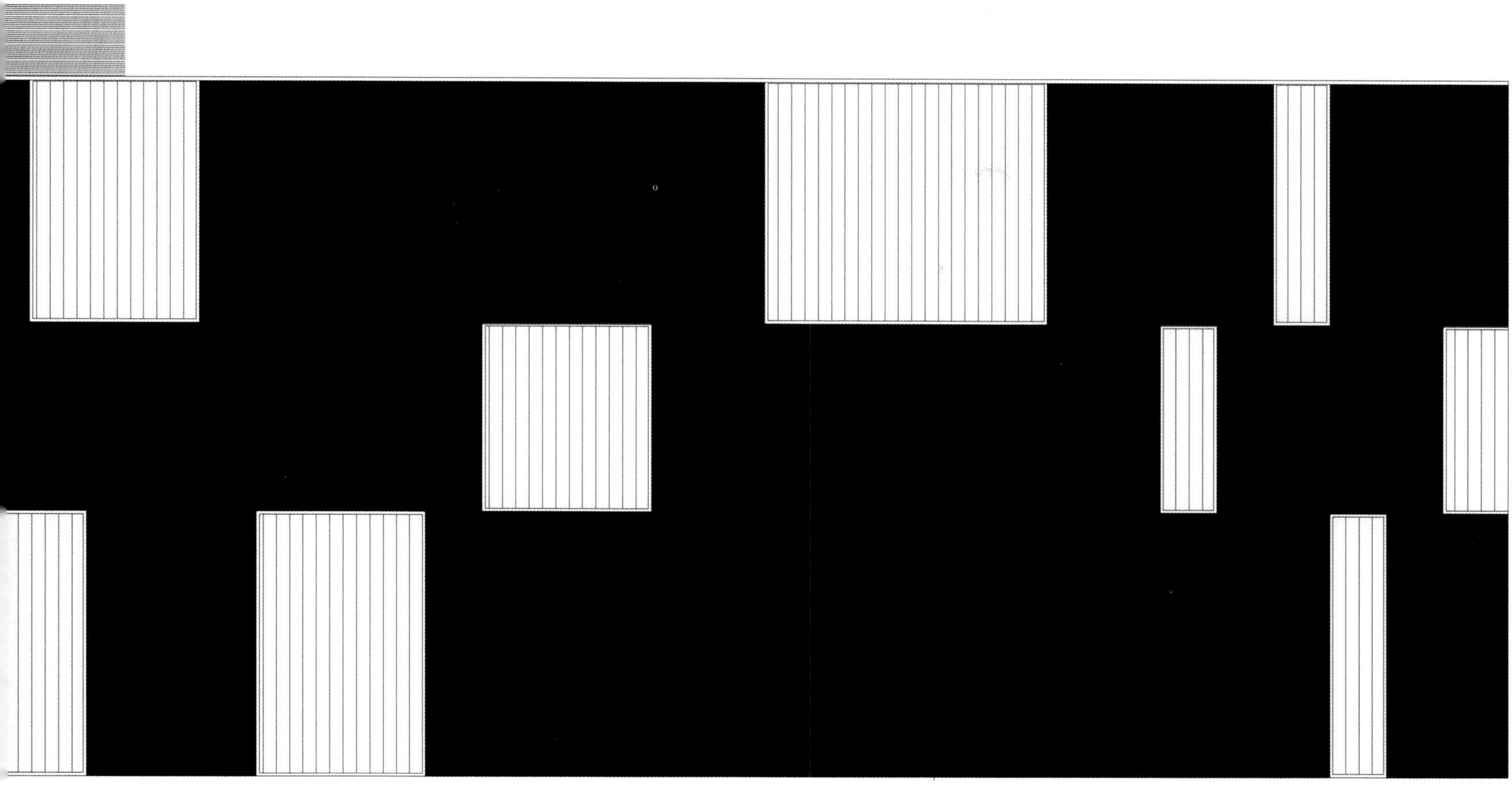

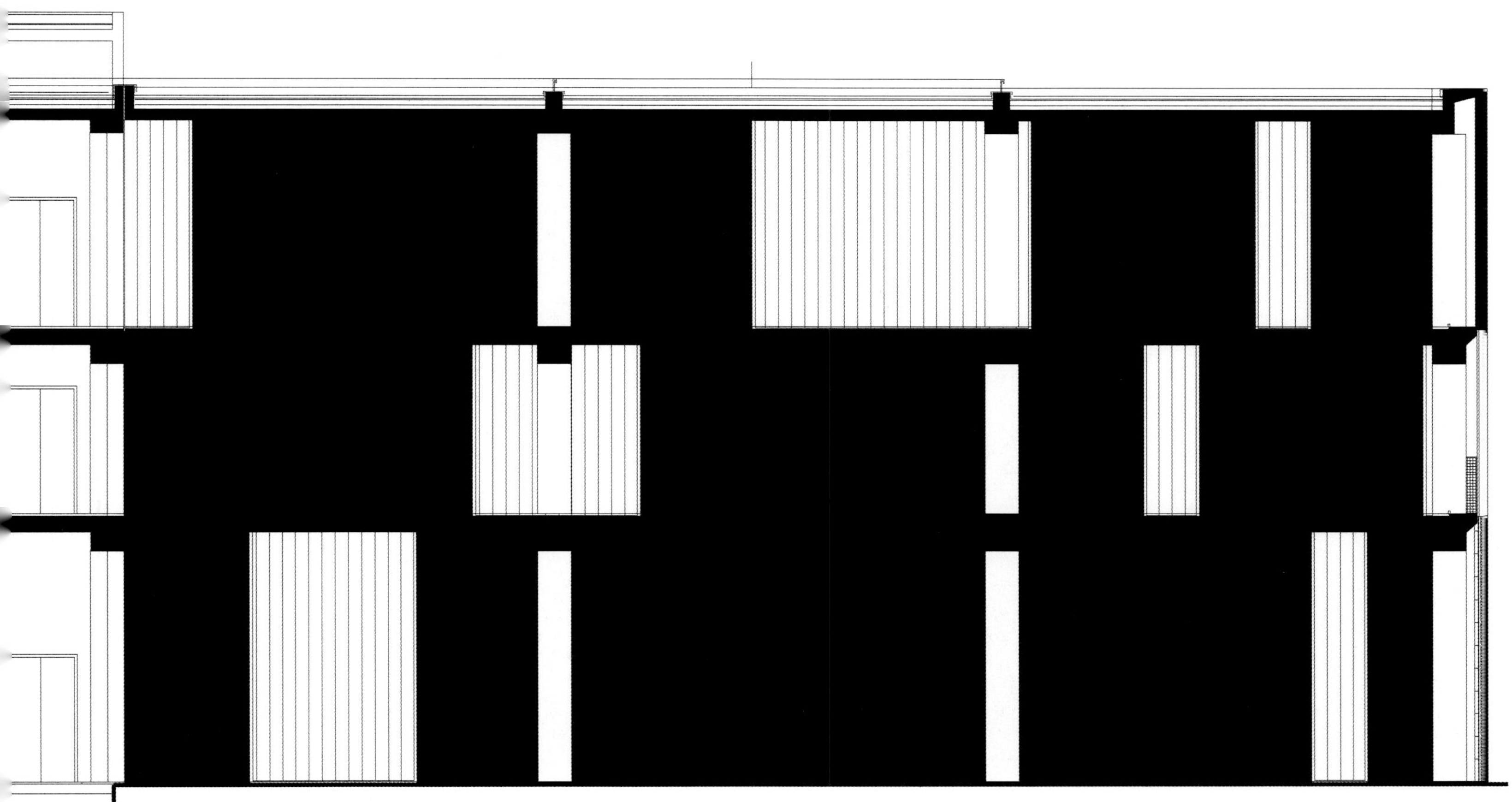

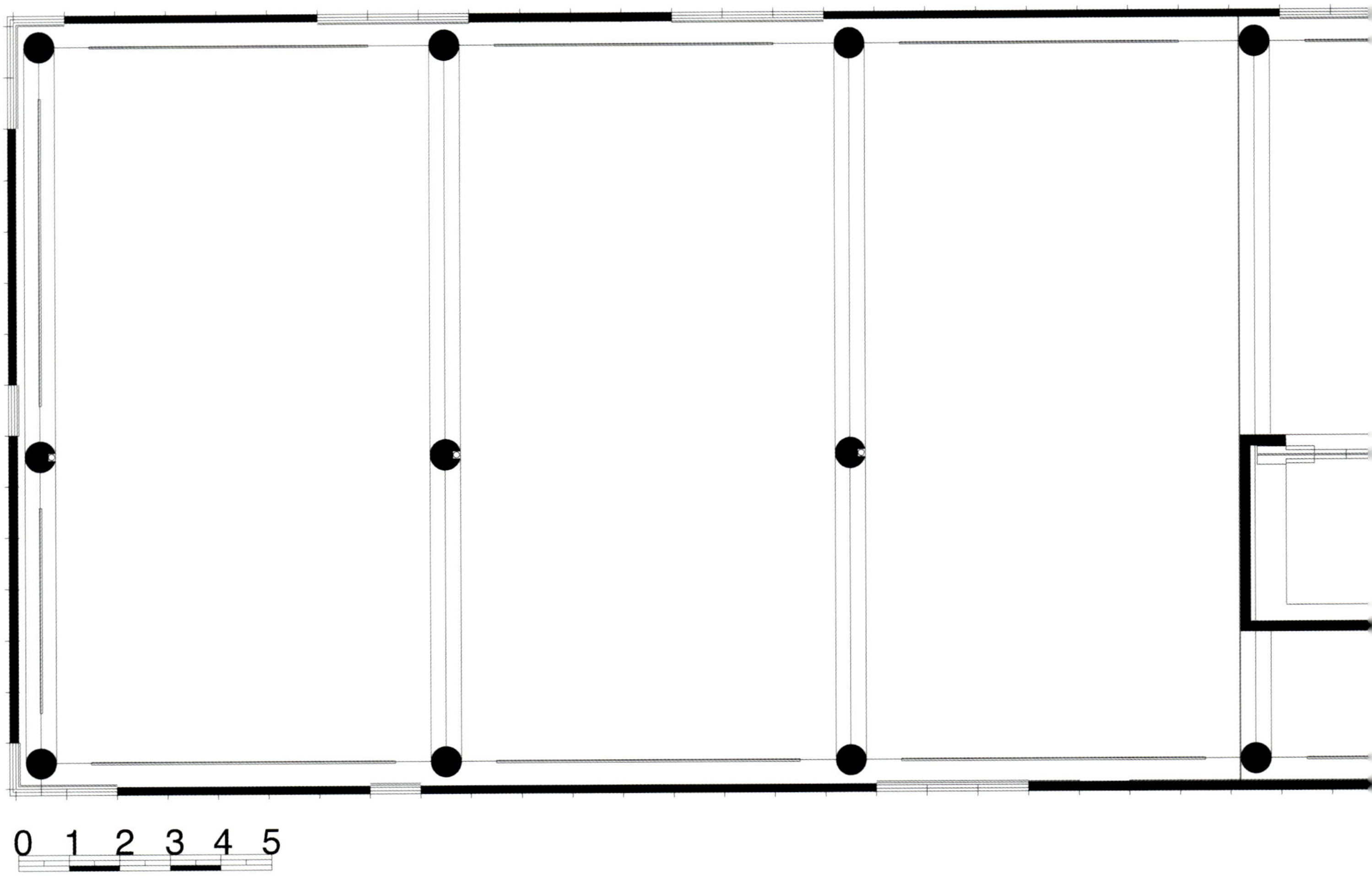

Floor plan

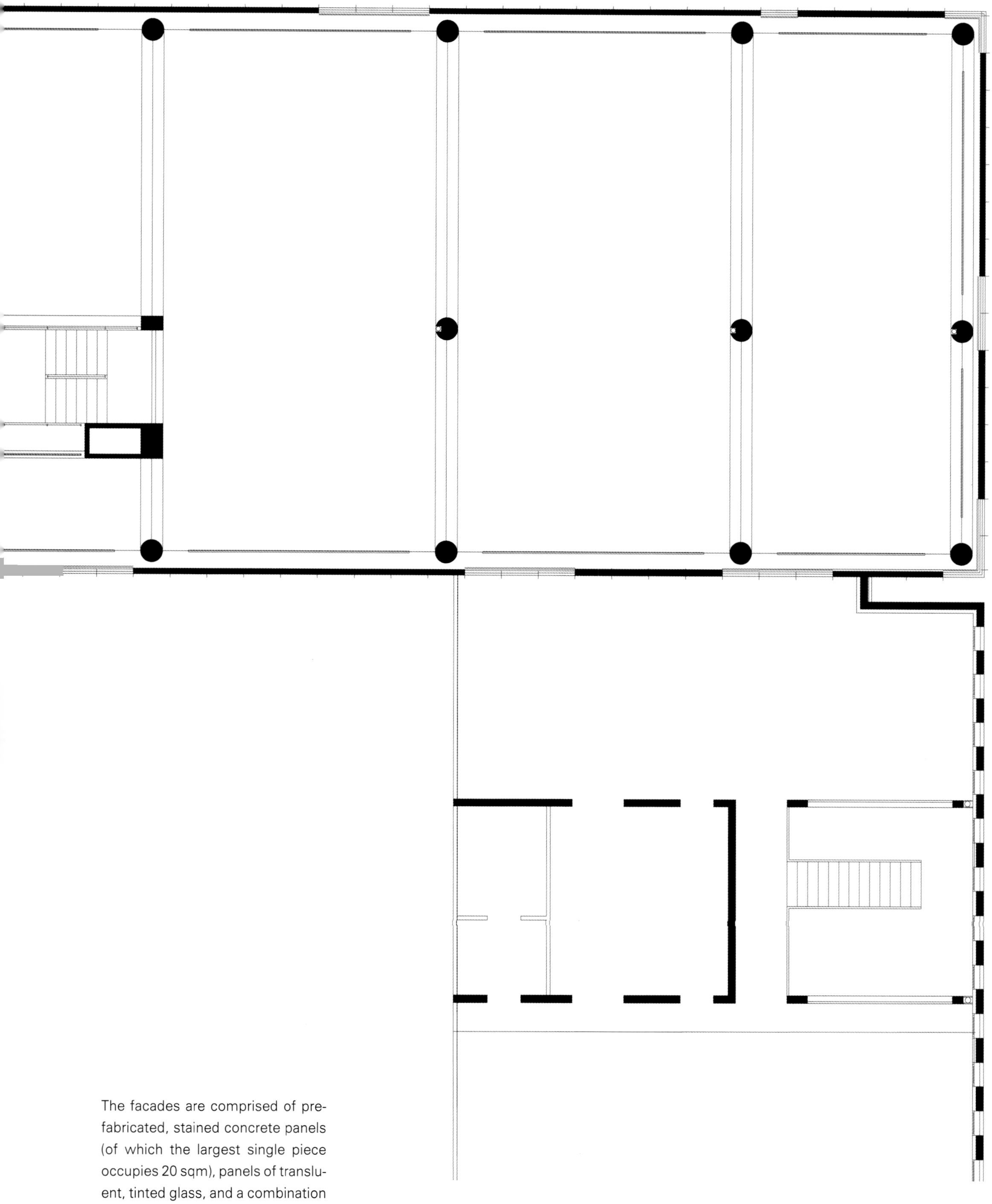

The facades are comprised of prefabricated, stained concrete panels (of which the largest single piece occupies 20 sqm), panels of translucent, tinted glass, and a combination of layers including metal panels, insulation, transparent polyester and an outer skin of black-tinted, perforated metal.

2
2

The new building is a hollow-bodied polygon formed by a concrete skeleton that releases the useful space from all loadbearing elements and whose dimension are stablished according to the movement of fork-lift trucks.

Bothe, Richter & Teherani

Company Building Tobias Grau

Rellingen, Germany

Photographs: Michael Wurzbach

The goal was to create a sophisticated architecure with means as simple as possible, such as timber glue truss beams. The original task consisted of the new erection of a commercial building for the company Tobias Grau with 30 employees at the time. The floor plan covers a prefab parts warehouse with areas for final assembly, delivery and shipping, as well as an office section for the commercial and design departments.

Now, with the completion of the second building phase, the building has been extended with a plan meant to respond to new demands of design, work flow and cost.

The first building phase consists of an elongated, oval edifice. A second, identical shape has been executed as the extension and connected to the first building by a two-story, almost square edifice.

The construction consists of timber glue truss beams spanning over 20 m, which were erected at a distance of 5 m and which support the external aluminum skin made of Alucobond plates. An inserted concrete table in the first building results in two-story use and at the same time serves as reinforcement. This concrete table is formed as a soffit in cross cap shape with a hog of 10 cm and stands on filigree concrete supports. On the upper floor the structure is completed by wooden rocking piers.

The building was executed completely free of support columns, as a large hall with timber glue truss beams was constructed accordingly.

As a result of the company's involvement in the lighting sector the opportunity arose for the company Tobias Grau —as owner of the building— to develop all lights anew, for this building in particular, and then for mass production.

Areas which feature new lighting elements include the conference room, the toilets, stairwell, cafeteria and work stations. The architecture of this company's building will thus partially function as an impulse for ideas and as a trigger for product and program development.

TOBIAS GRAU

Upper floor plan

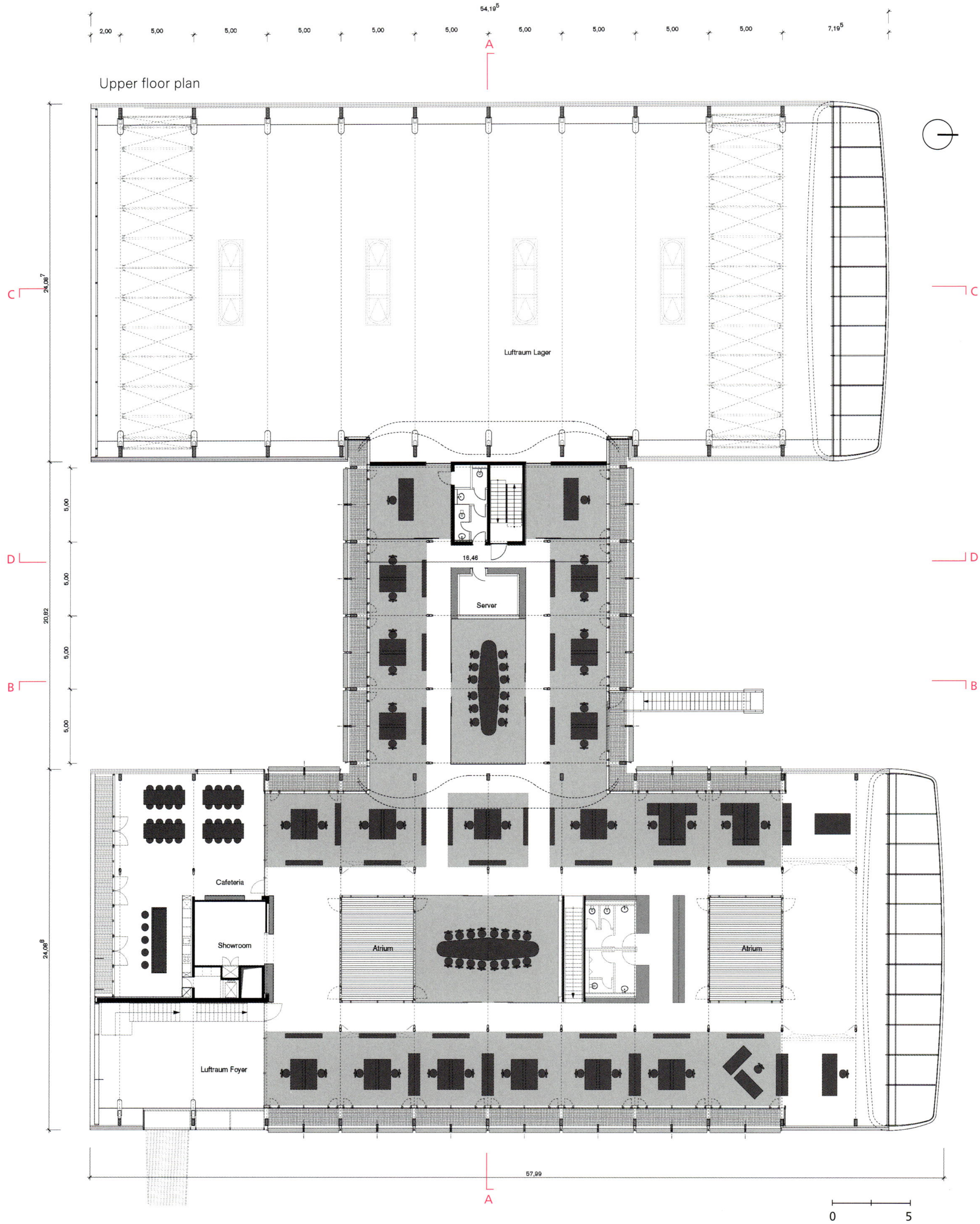

Ground floor plan

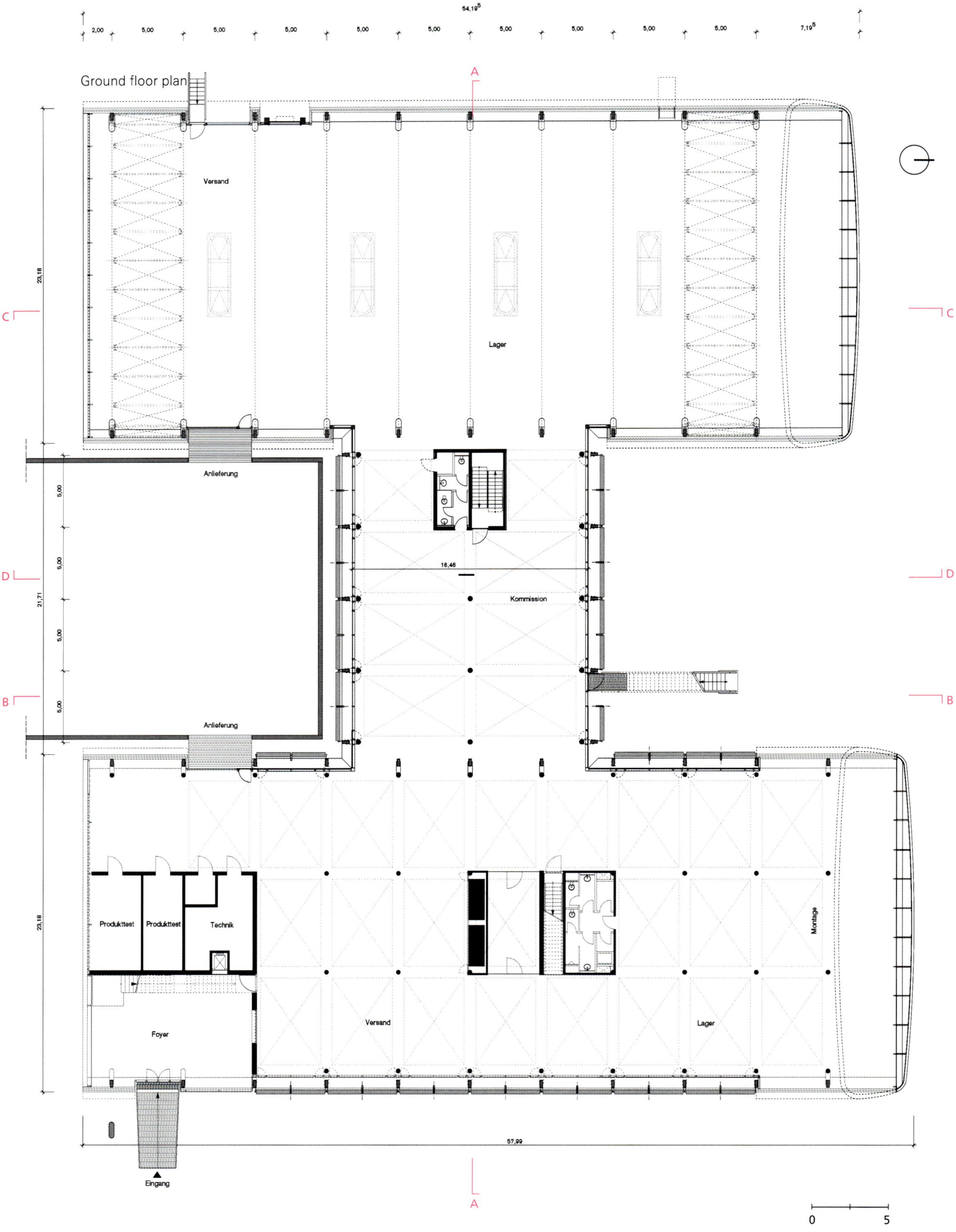

The building consists of timber glue truss beams spanning over 20 m, which were erected at a distance of 5 m, and which support the external aluminum covering, made of Alucobond sheets. An inserted cross-shaped concrete table in the first building results in two-story use and at the same time serves as reinforcement.

North elevation

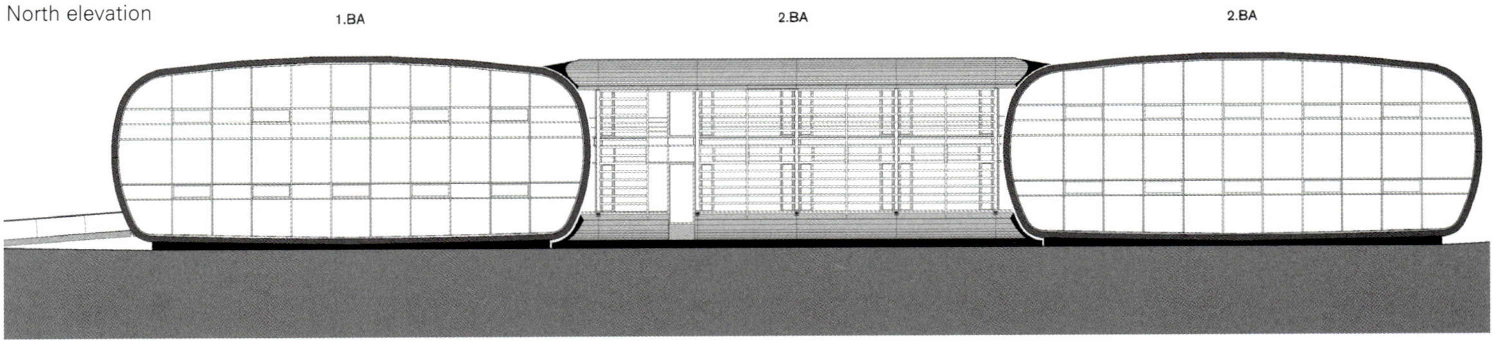

South elevation

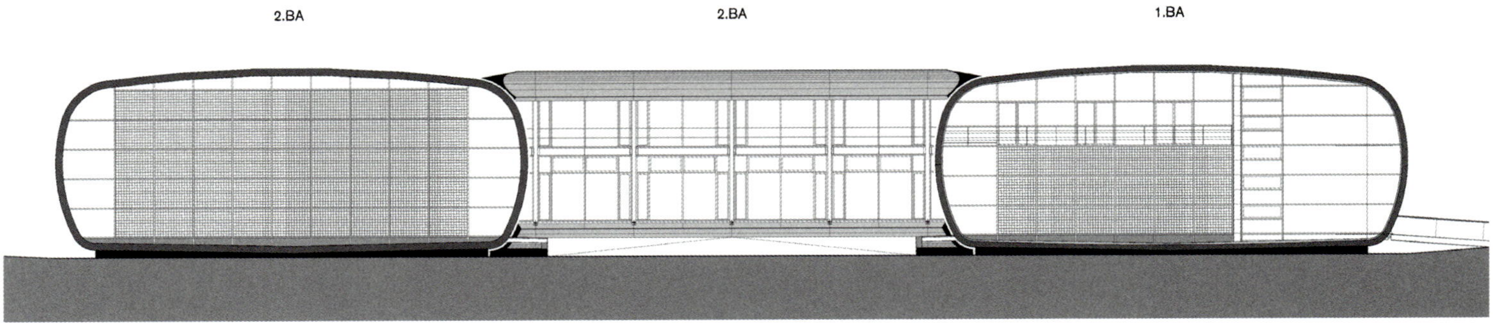

West elevation

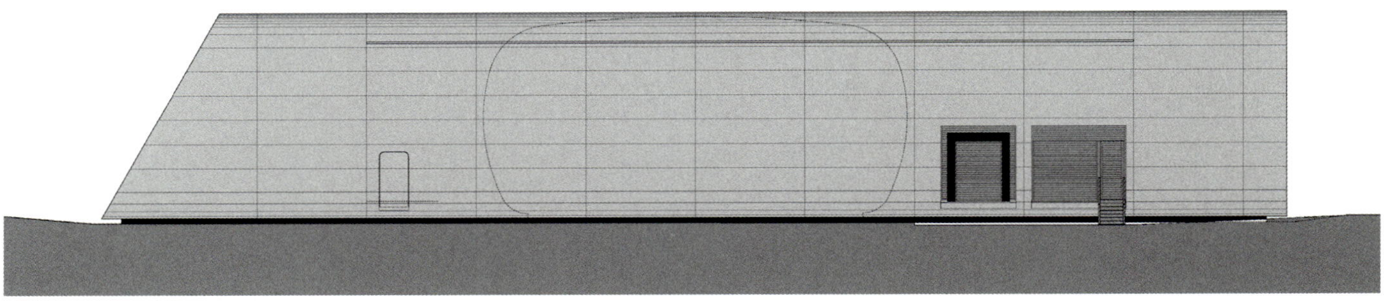

Section AA

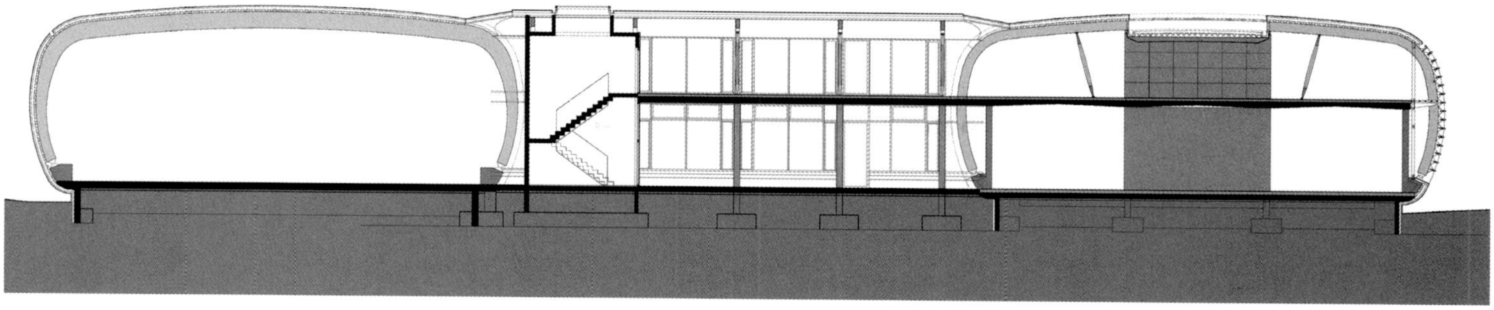

Section BB

Section CC

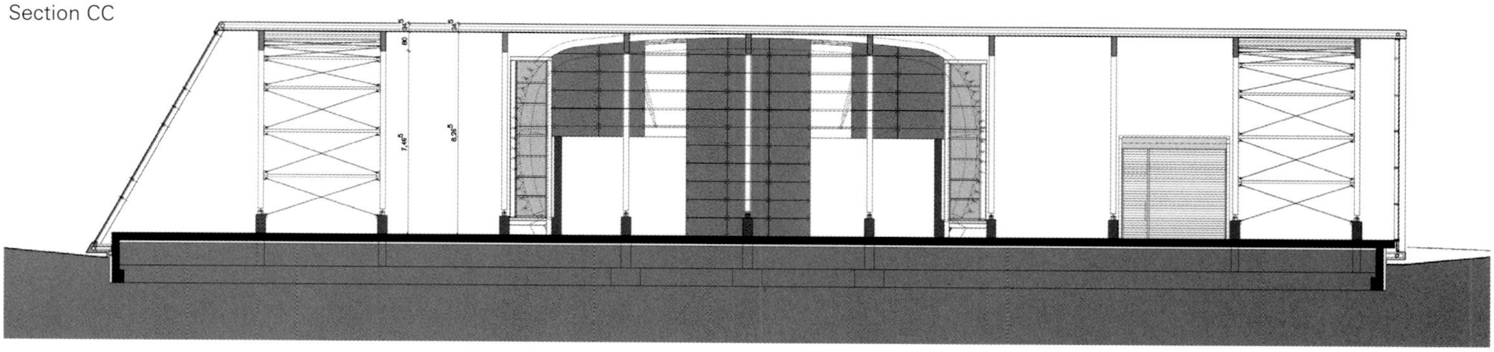

Section DD

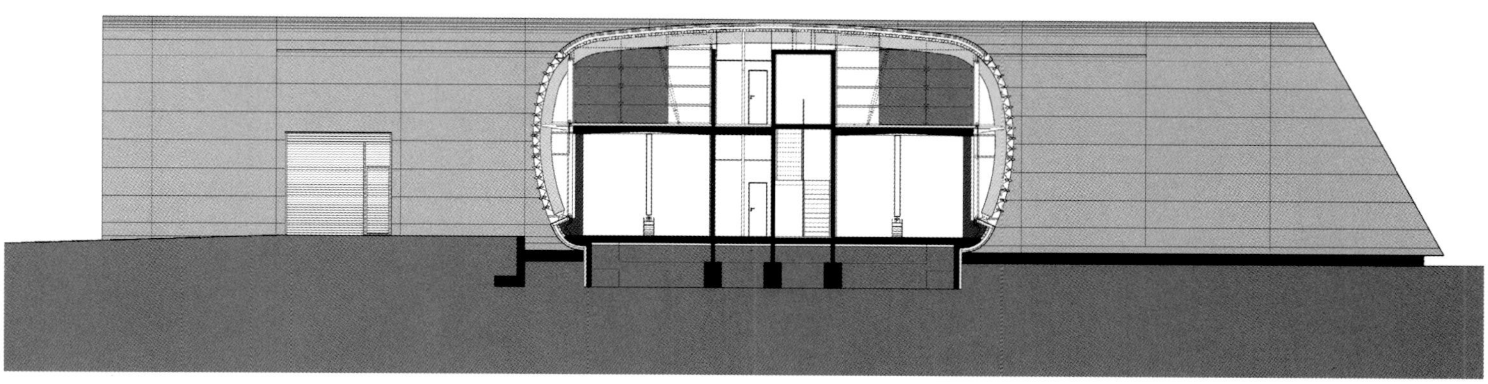

The building was executed completely free of support columns, as a large hall with timber glue truss beams was constructed accordingly. On the upper floor, the structure is completed by wooden rocking piers. As a result of the company's involvement in the lighting sector, all lights in this project were developed anew.

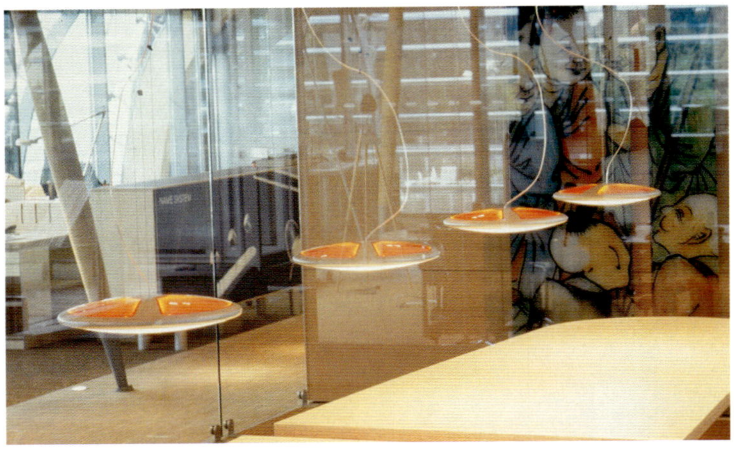

Foster and Partners

The Design Center

Essen, Germany

Photographs: Nigel Young / Foster and Partners

This former powerhouse in an Essen coal-mining complex forms part of an extraordinary group of early 20th century buildings, all with red-brown painted exposed steel I-beams, and an infill of industrial glazing and red brick. These magnificent structures, with their towering chimneys and vast halls, are the cathedrals of the industrial age. The powerhouse has an inner hall of colossal proportions, as impressive as any Gothic edifice. The challenge was to adapt this building to create a new home for the German Design Centre, without fundamentally altering its character.

The building's façade was restored and a number of later additions were removed to reveal its original form. Inside, the heavy industrial feel of the building has been conserved. One of the large boilers was preserved intact as an example of 1930's technology. The remaining four boilers were removed and replaced by independently supported galleries, which are articulated from the existing structure as 'boxes within a box', their lightness juxtaposed with the heaviness of the original fabric. A simple concrete cube contains further exhibition space as well as conference rooms with views of the surrounding complex and the landscape. Visitors enter via the dramatic central hall where the rusty steel structure and exposed brick walls are still visible.

Together with associated offices, the Design Centre houses a combination of temporary and permanent exhibitions (everything from cars to electrical appliances) that are constantly updated and require highly flexible galleries. The different exhibition areas and the interaction of old and new architecture, create a varied backdrop for the location of exhibits, while the changing nature of the exhibitions themselves adds a further dynamic element to this relationship.

Ground floor plan

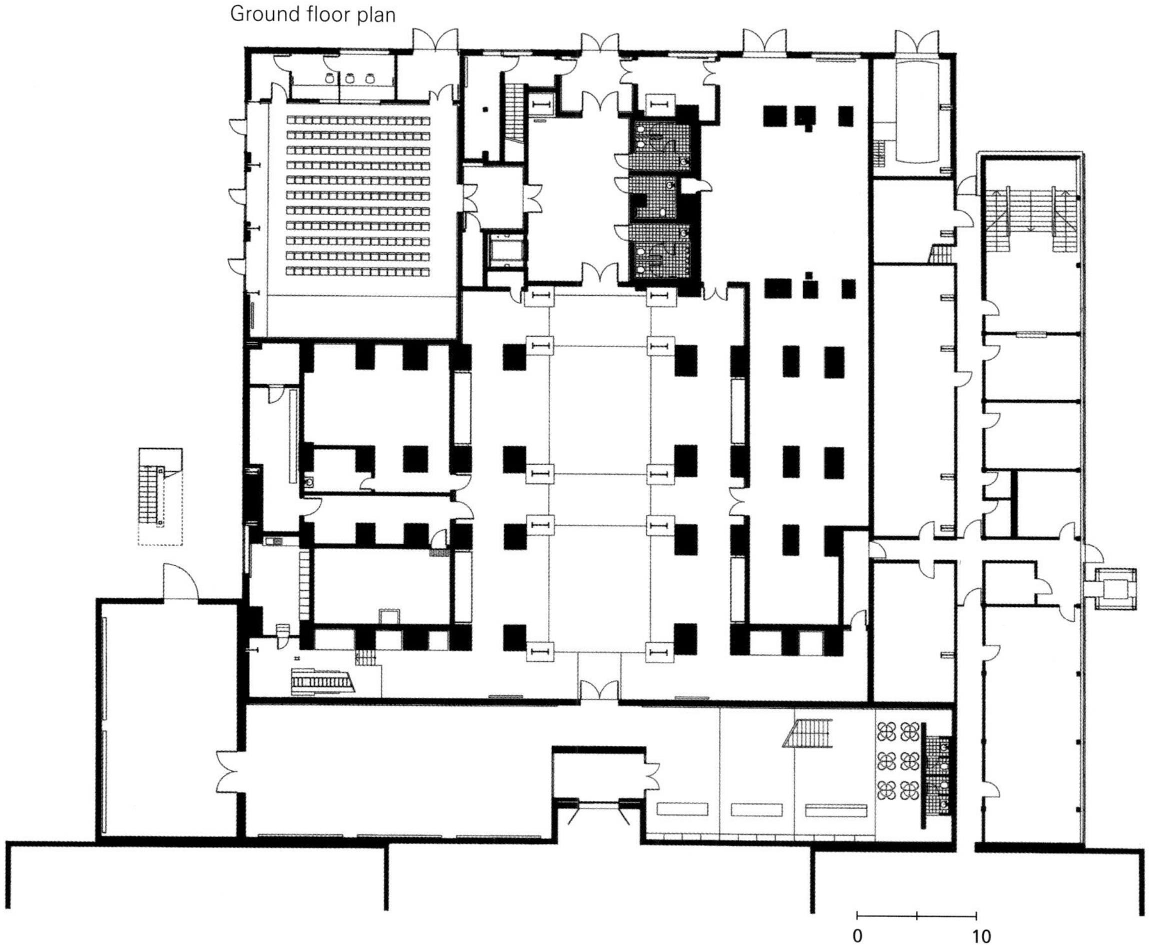

Second floor plan

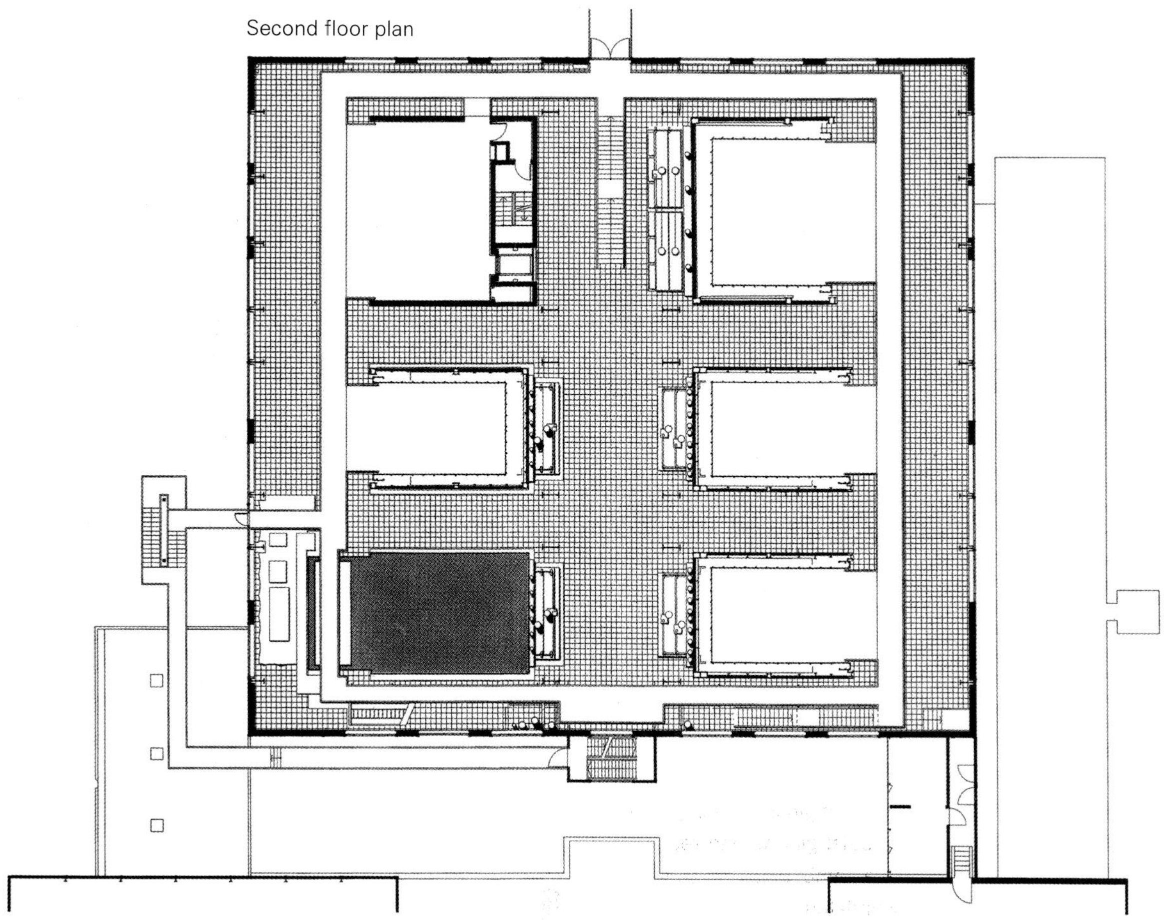

The final project in this powerhouse conserves the original red-brown painted exposed steel I-beams, and an infill of industrial glazing and red brick. Highly flexible galleries have been provided for a combination of temporary and permanent exhibitions that are constantly updated.

Fourth floor plan

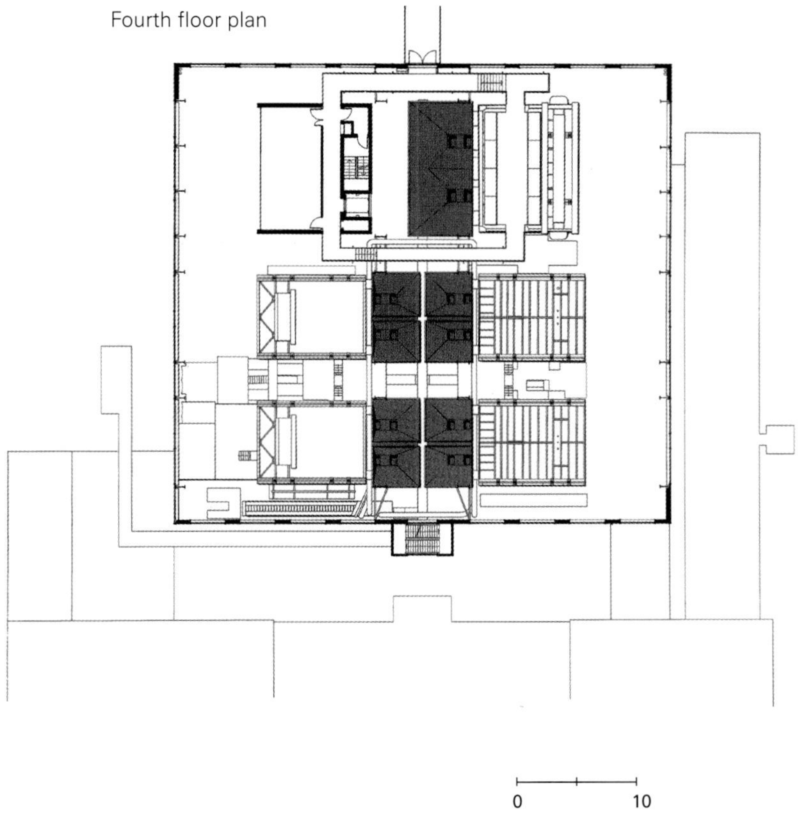

Fifth floor plan

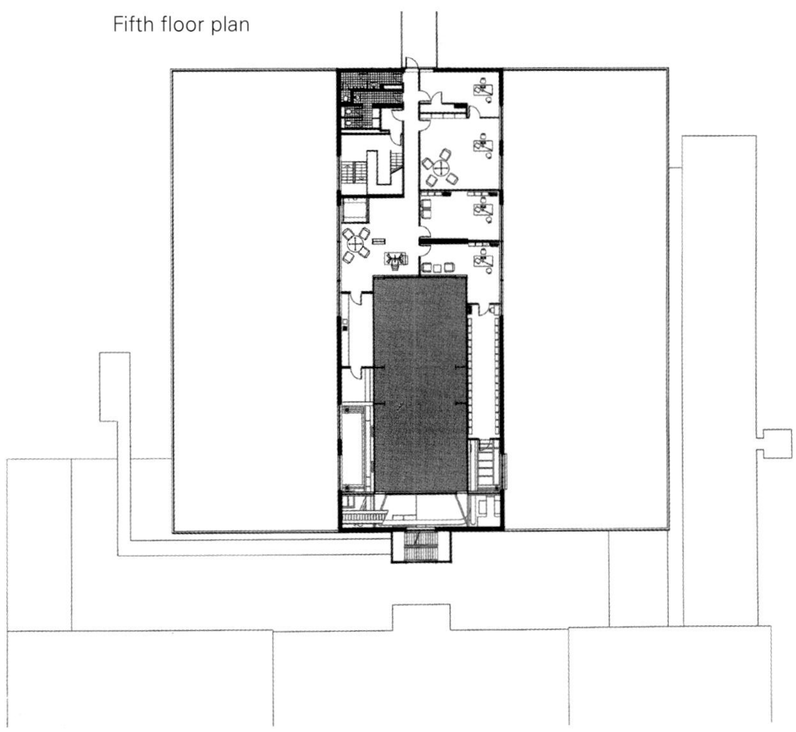

Longitudinal section

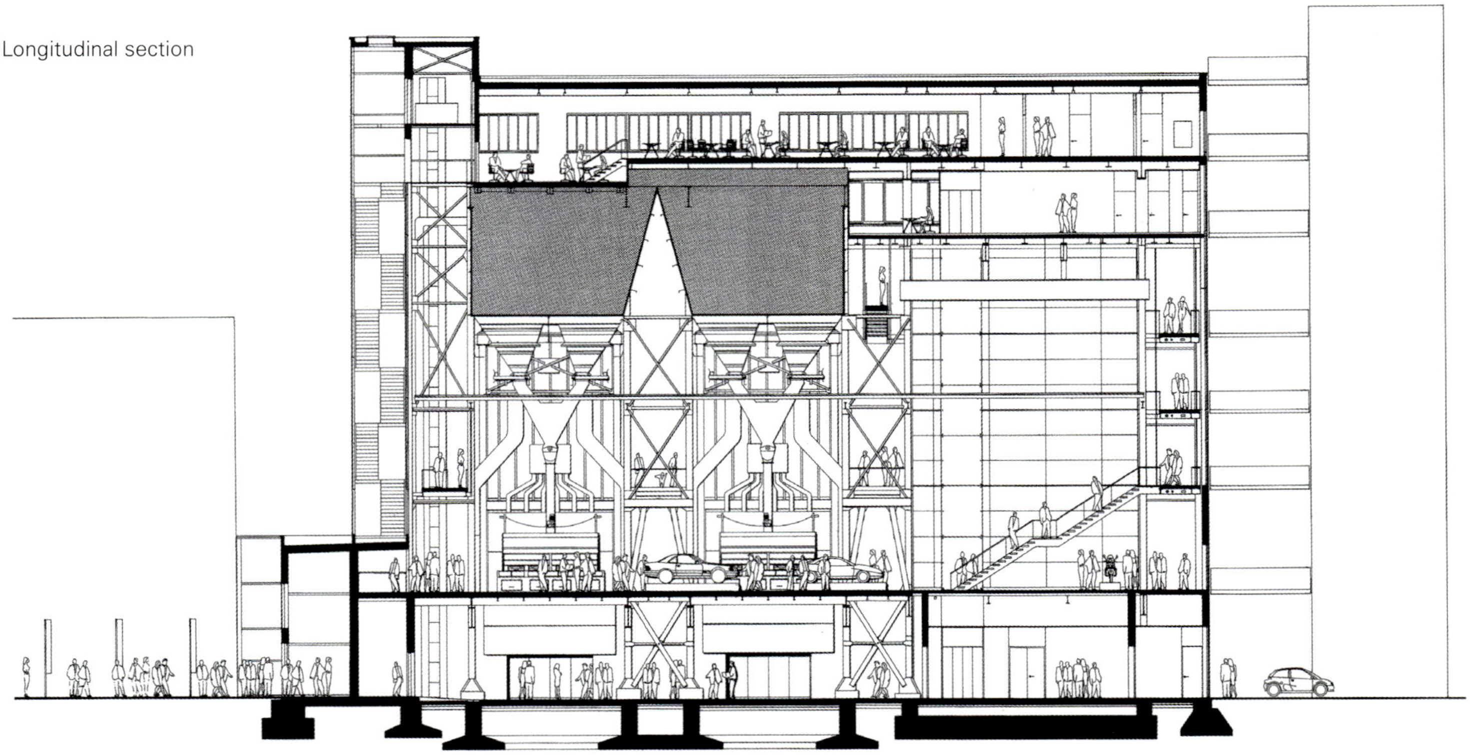

The heavy industrial feel of the building has been conserved by leaving much of the listed structure exposed and un-restored. Four of the five boilers were gutted and two floors of galleries inserted. They provide a blank canvas for the exhibitions whilst still revealing the upper parts of the boilers. These galleries are accessed from the outer edges and are linked to each other by a walkway with glass balustrades.

Cross section

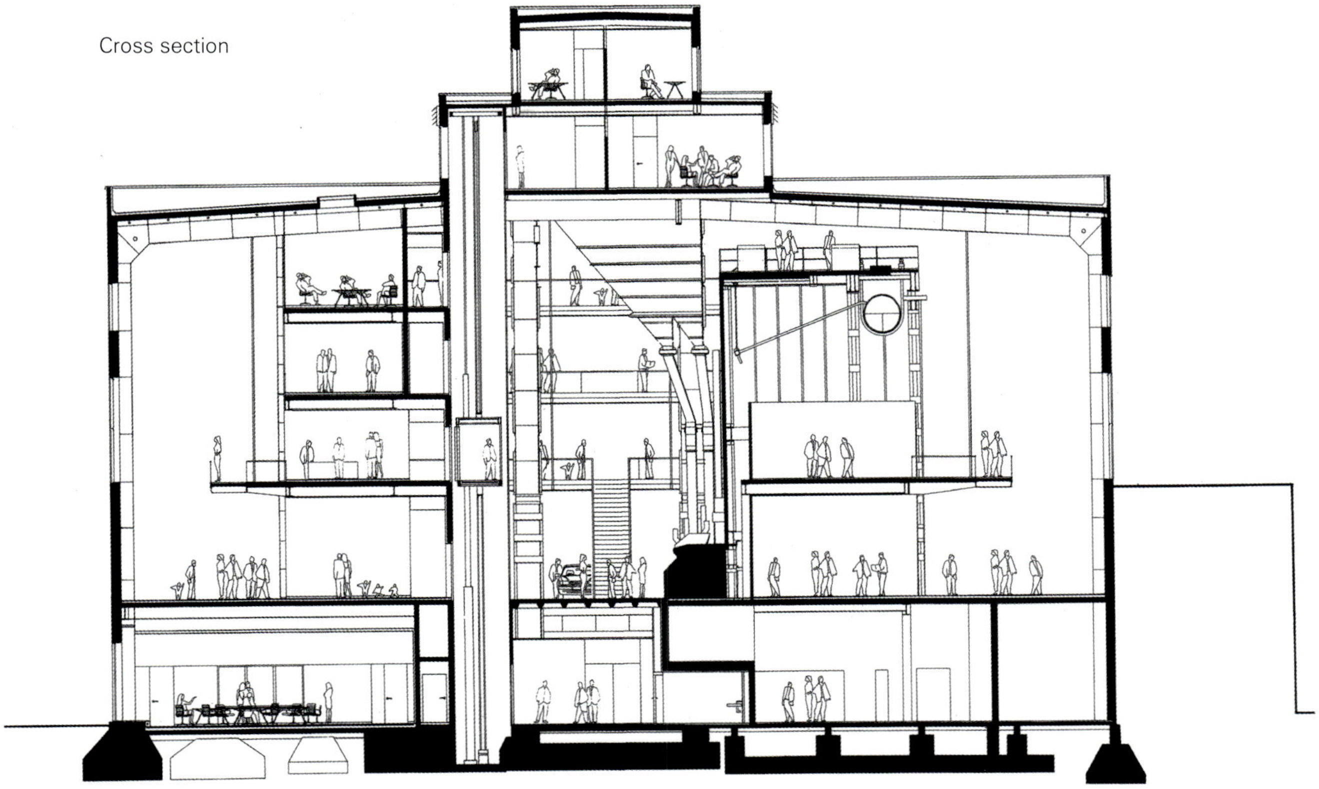

Richard Hywel Evans Architecture & Design Ltd

Cellular Operations Headquarters

Swindon, UK

Photographs: John N. MacLean & Tim Soar

Cellular Operation's call center in Swindon has been celebrated in the national Press and Television as 'the happiest work place in Britain'. One of the company's objectives when commissioning the building to Richard Hywel Evans was to create an excellent working environment and a striking corporate identity. Another consideration was to design a flexible office space, paramount in a call center environment.

The end result was 4500 sqm of column-free space in a building that combines high technology with international-style serenity. The main structure of the building is a mass concrete frame which stores heat in winter and is cool in summer. Large sections of the south-facing façade are glazed in Okalux insulated glass, allowing daylight to penetrate with little solar gain.

Service and plant rooms are located on the second floor, behind an elegantly louvered 'tower', looking out across the main two-story workspace. Behind the service area, which also houses the Perspex lifts, are the main conference rooms and offices. These break out of the traditional rectilinear partitions in favor of an organic approach that reflects the building's exterior.

However, it is the great bulbous glass extension, supported on slim steel tubular columns, that is the highlight of the building. The spectacular curved façade brings daylight deep into the building and functions as the key element in a simple ventilation system, drawing cool air off the adjoining lake, which is then dispersed through a network of vents running through the floor to the south area of the building.

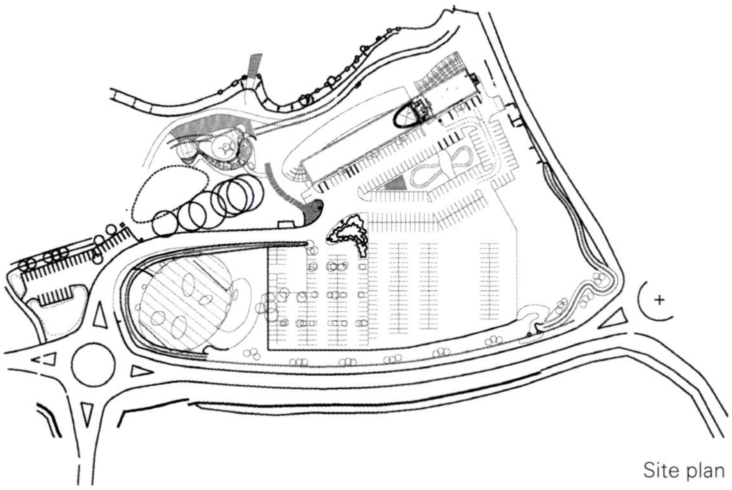

Site plan

Ground floor plan

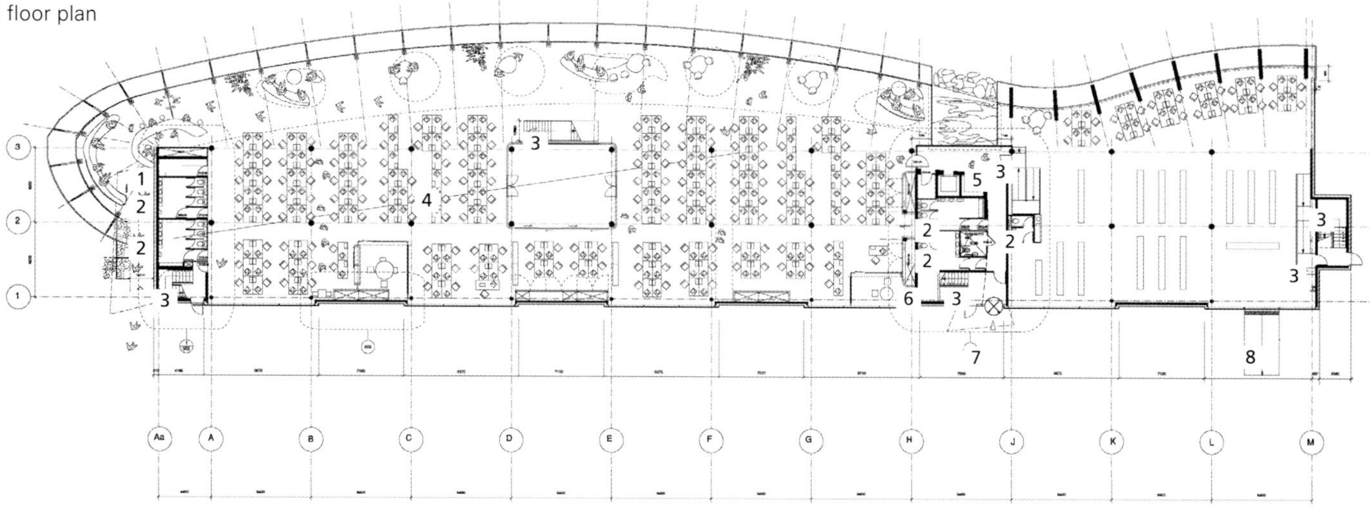

North elevation

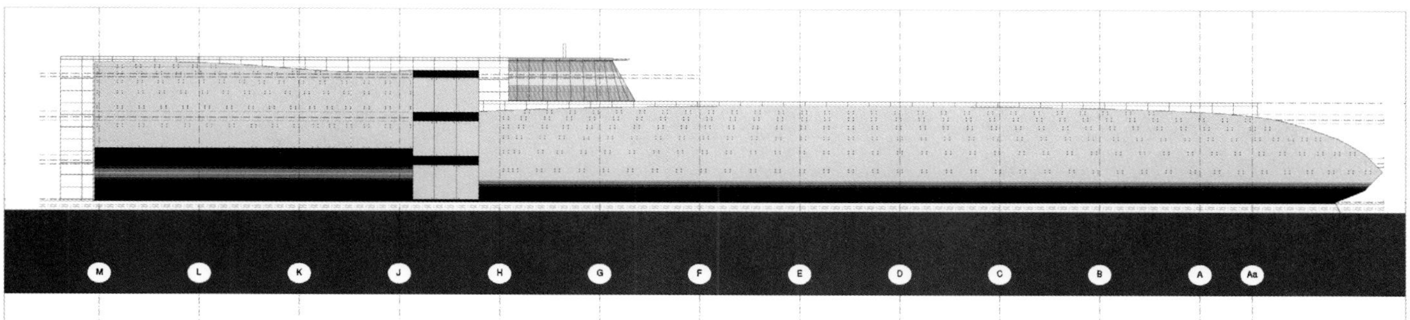

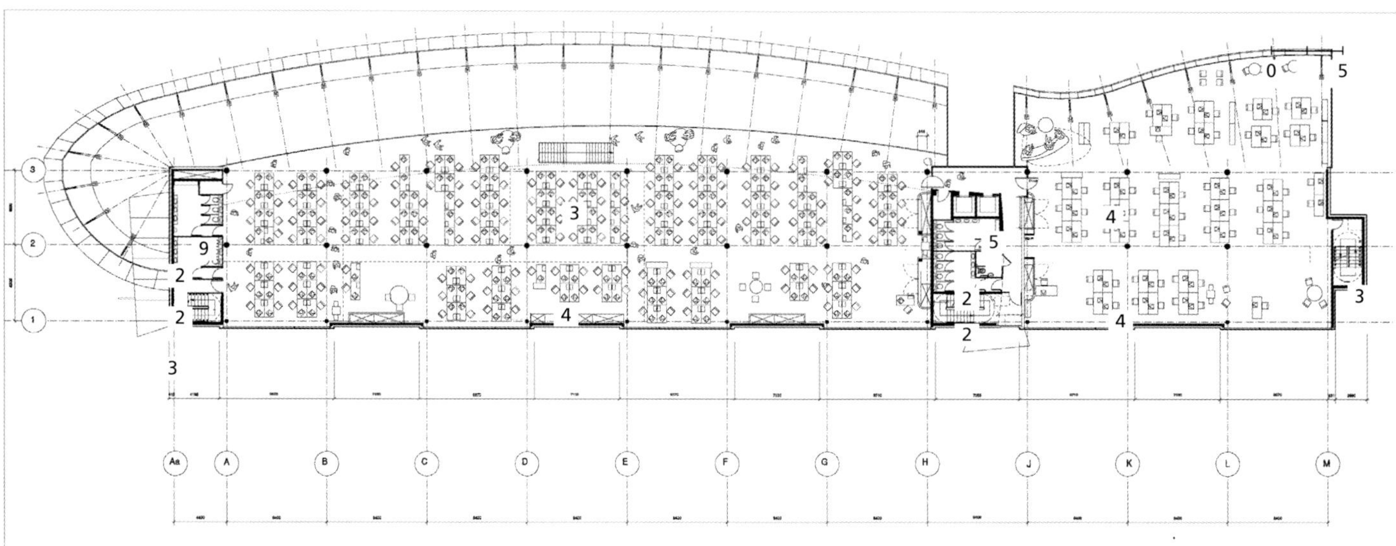

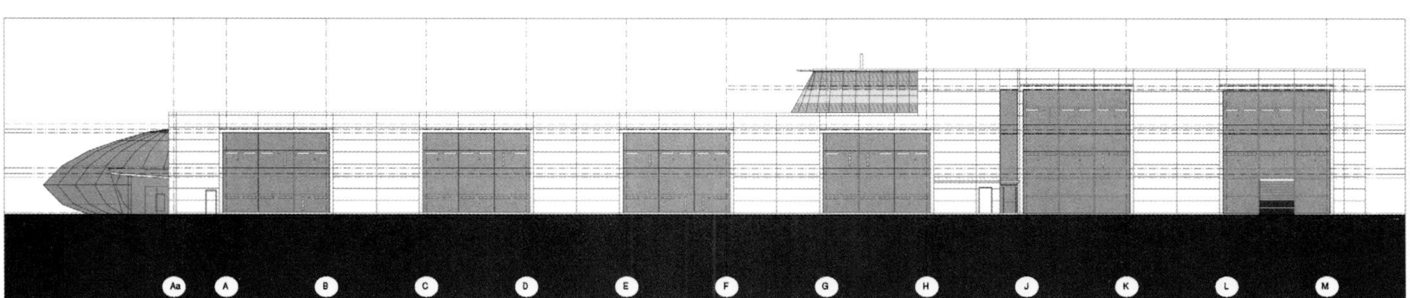

South elevation

1. Kitchen
2. Bathroom
3. Staircase
4. Office
5. Hall
6. Water tank
7. Staff entrance
8. Storeage bay entrance
9. Store

Second floor plan

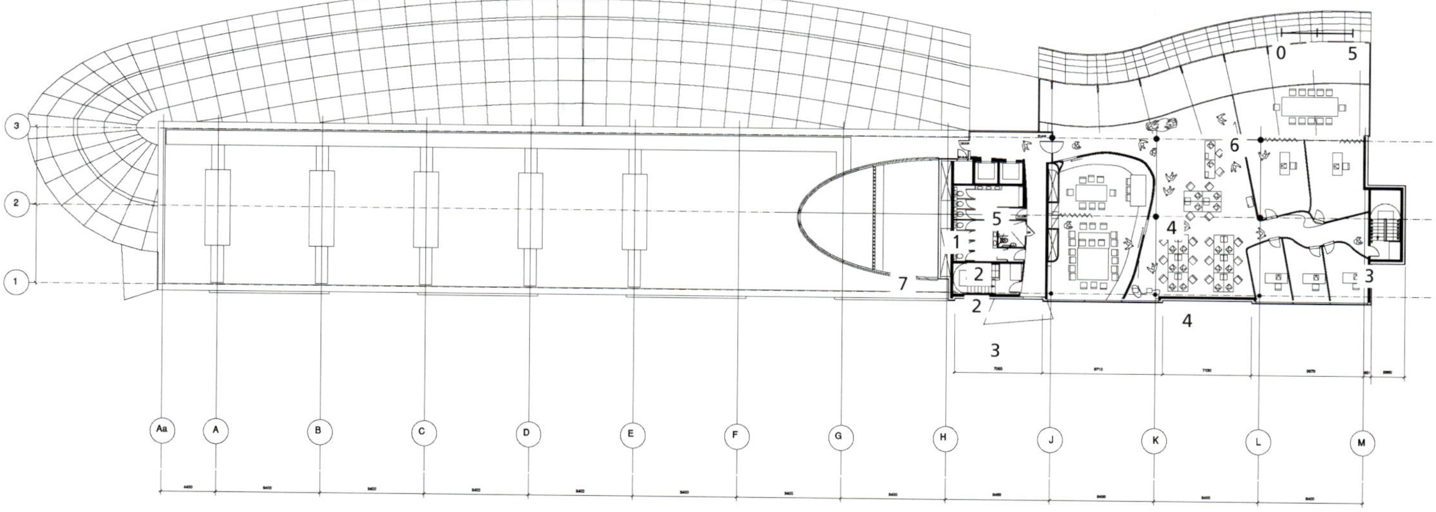

East elevation

West elevation

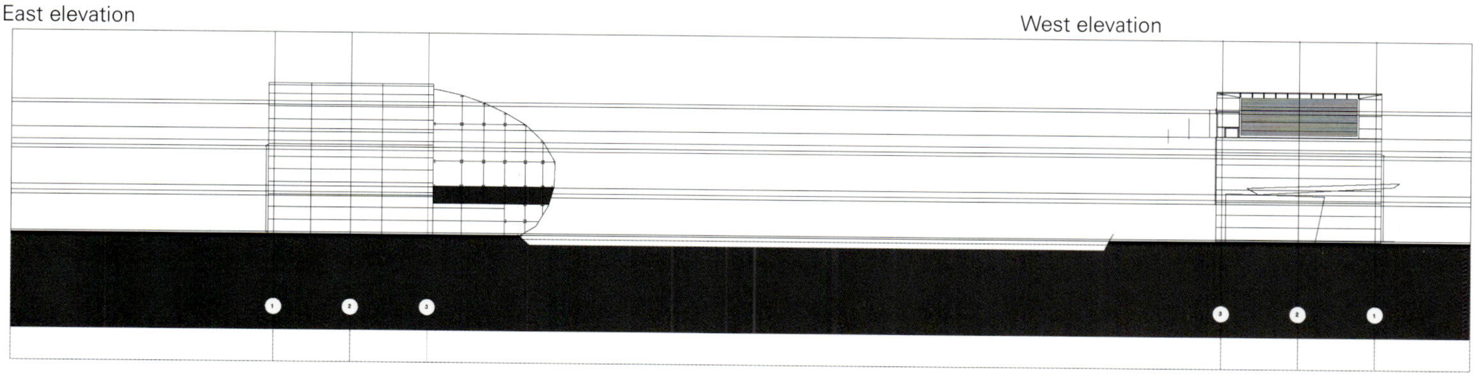

1. Motor room
2. Bathroom
3. Staircase
4. Office
5. Lobby
6. Void
7. Plant room

The internal staircase in the IT room, pictured right, is formed by poured concrete and underlit lozenge-shaped glass treads. The building is wired for the latest IT, and all systems feed into a central 'brain', a spectacular glass-walled room at the heart of the building.

Cross section

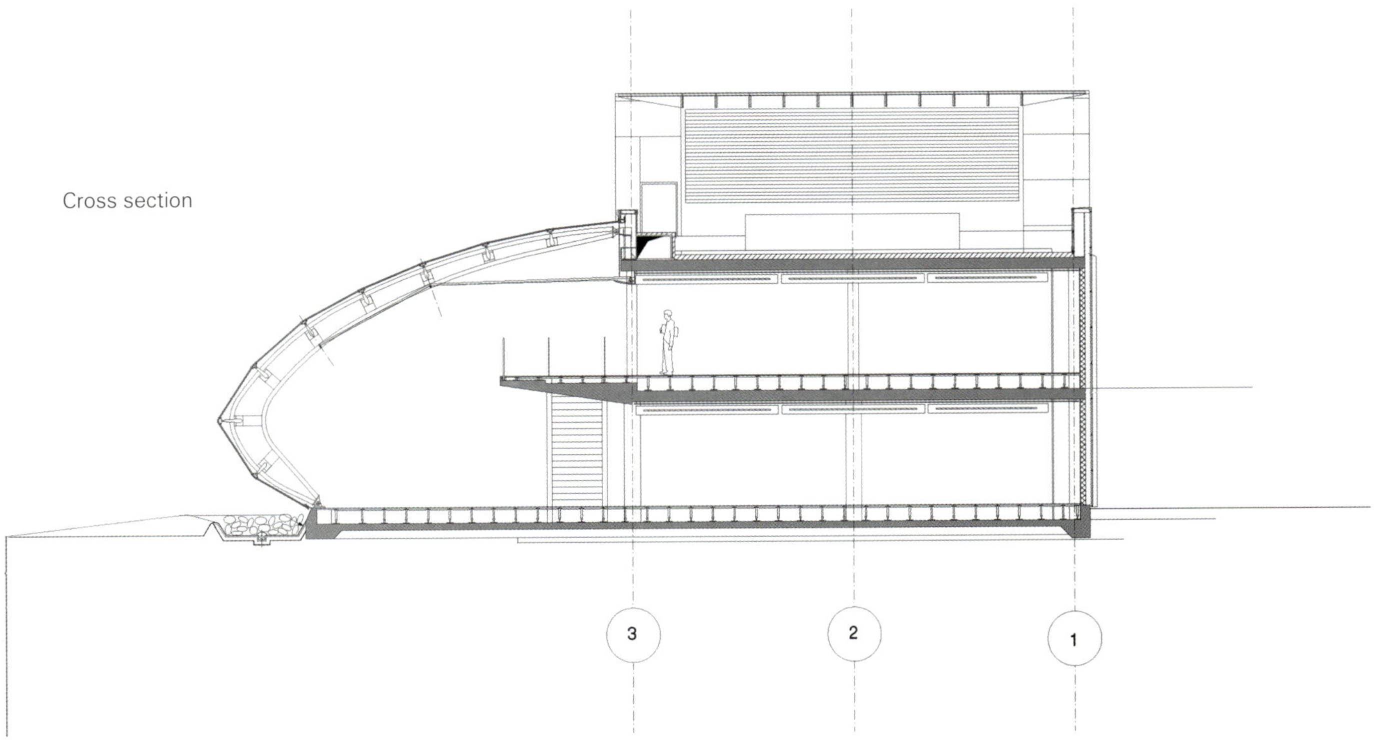

A curved coachbuilt riveted aluminum reception desk in the glazed area doubles as a juice and cappuccino bar, effectively mediating between public and private space. The company was keen to bring back a symbol of British '70's nostalgia: the uniformed tea lady with her trolley.
Dedicating particular attention to a traditionally neglected area of office design, the architect has designed mosaic-clad WCs that feature cast hand basins and electronic taps.

Herzog & De Meuron

Ricola Marketing Building

Laufen, Switzerland

Photographs: Margherita Spiluttini

The new building is located in the middle of a group of small-scale village buildings without any remarkable architectural qualities. The surrounding gardens, however, with their hedges and trees, offered a wonderful environment for a transparent architecture relating the interior and exterior spaces.

Instead of a multi-story volume that would dominate the architectural landscape, the architects chose a low polygonal installation that fits into Ricola's garden area like a pavilion, an architecture in which outer form and geometry did not immediately reveal themselves. One which, thanks to its turned-back facades, dissolves into single pieces. Each piece has the distinct characteristic of either surrounding, reflecting, or projecting far into the building's interior a special site in the garden.

The deep cantilevered roof can be understood as a symbol for the strategy utilised here of melting nature and architecture. Roof beams are made of a special plastic having variable flexibility for assimilation to changing temperatures and the changing weight of rain or snow. Plants, woven among the beams, form a hybrid natural-artificial construction which lends the building an ever-changing appearance in accordance with the passing seasons. Ivy provides for basic green throughout the year between the beams. The leaves of other plants -wild vine, for example- are only visible during the summer months and help to prevent too much sun from penetrating the glass facades.

Inside, the building is planned as a single, cohesive, open space that offers, for the most part, a transparent office landscape on two floors. The large staircase in the middle of the building is simultaneously a connecting element, meeting place and auditorium. The building is equipped with facades of pure glass throughout, with a few wall-height sliding doors. Thus actual spatial delineation is not static and may be changed according to need. Curtains mounted on three parallel runners allow users differing variables as to color, transparency and view.

Like its landscape design, the building's textile equipment is indivisibly bound to its architectural concept. In no way is it a more or less coincidental decorative element. Its conception demanded early and close co-operation with the landscape architect Kienast Vogt and with artists Rosmarie Trockel and Adrian Schiess.

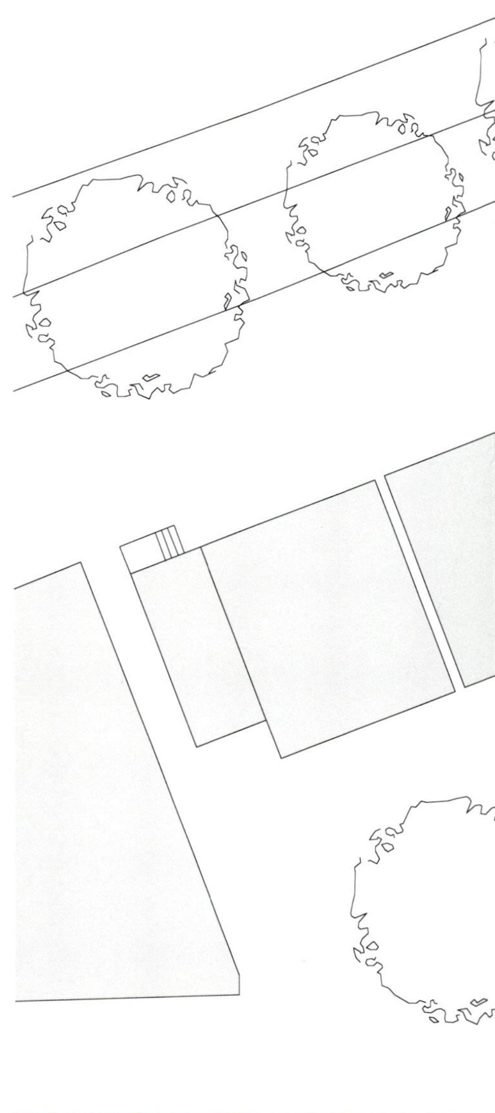

Ground floor plan

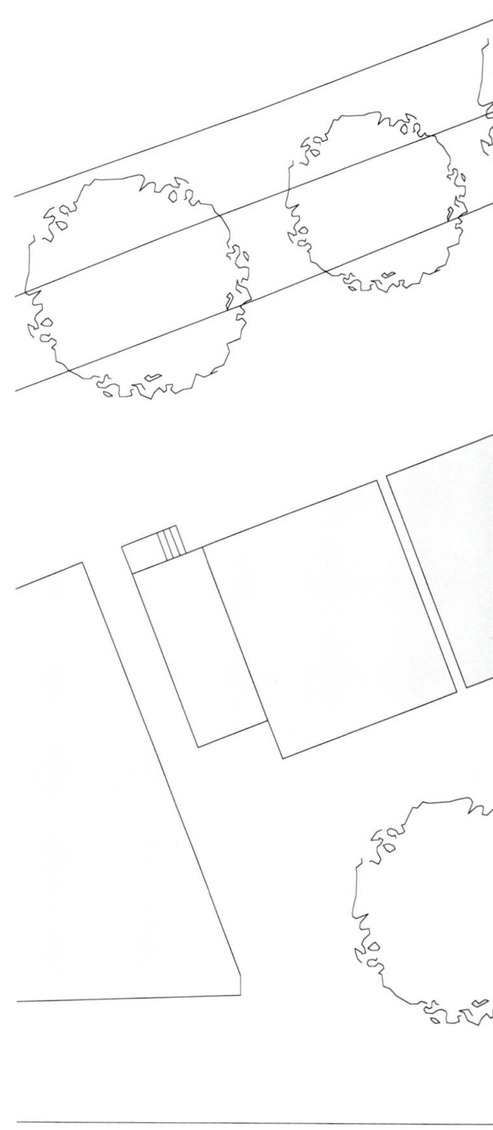

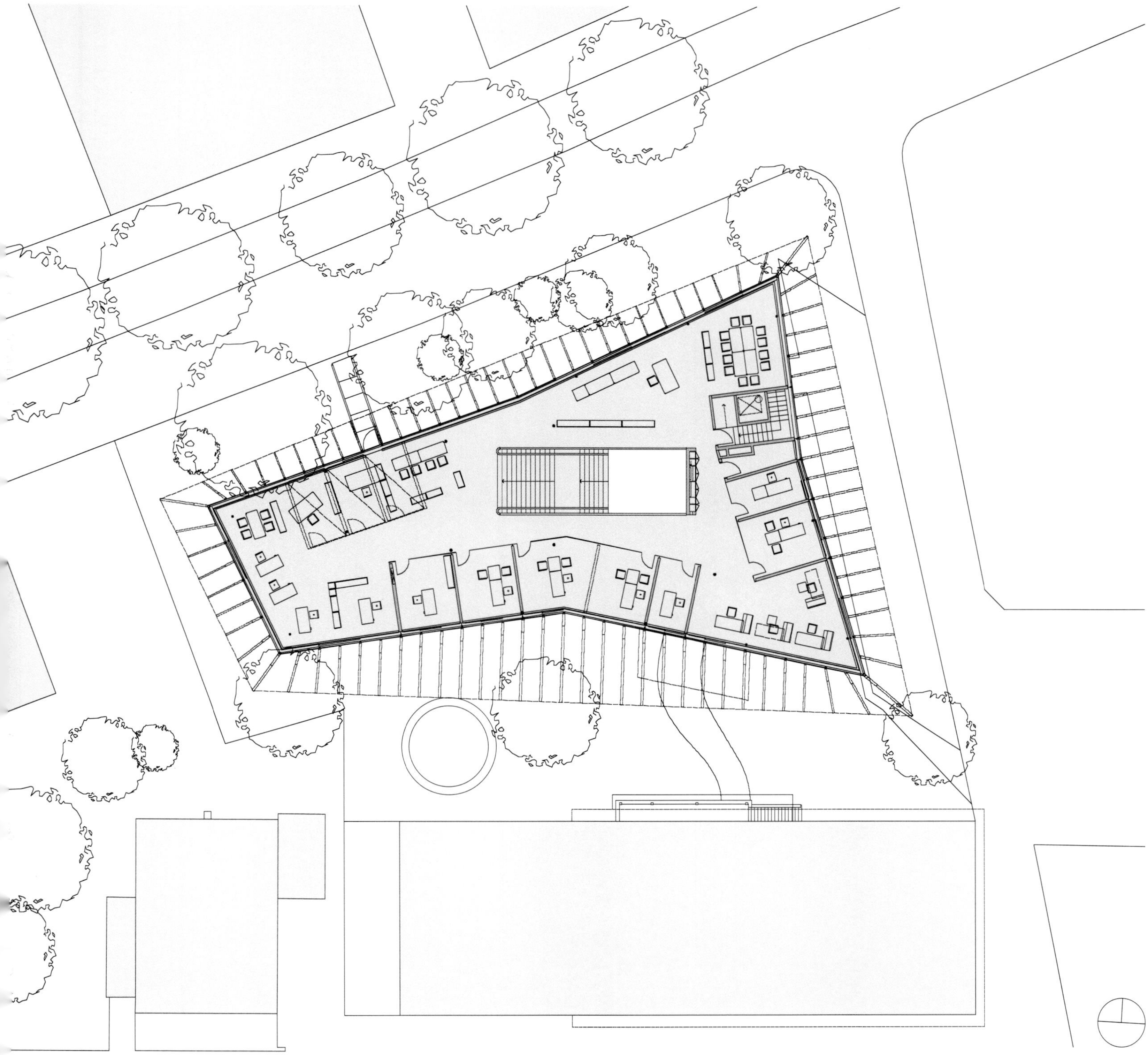

First floor plan

West elevation

Longitudinal section

South elevation

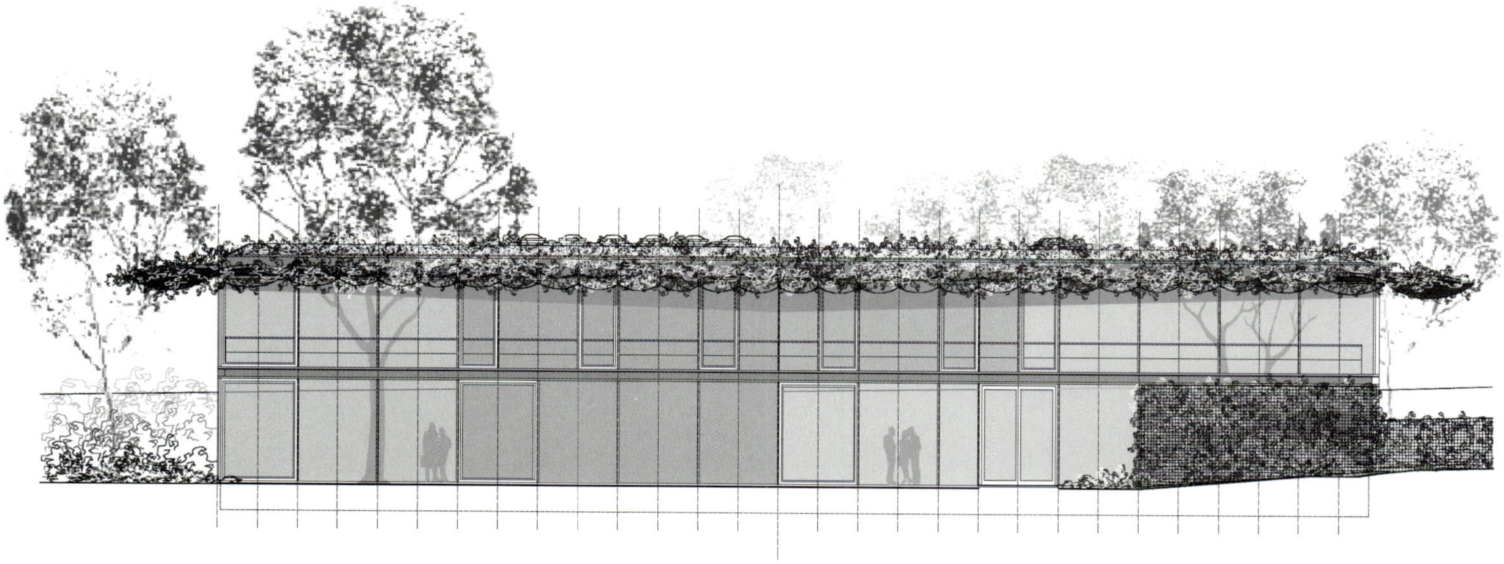

Cross-section

The interior of the building is conceived as a transparent space, with glass divisions that evoke the glazed facade and favor communication between the members of the firm. Three layers of curtains mounted on parallel rails allow an interplay between variables of color, light and views.